Science, God, Universe

and

PRAYER

Dr. Vijay Mohan Das

Science, God, Universe and Prayer
Copyright © 2014 Dr. Vijay Mohan Das

Requests for permission should be addressed to
Dr. Vijay Mohan Das, Subordinate to Almighty B.B.B
University of God , Das Nursing Home , Fatehgarh, Farrukhabad U.P. India , 209601.
(dasvijaymohan1@gmail.com, vijaydas@sancharnet.in)

Contents

Contents	3
Dedication	5
Acknowledgement	7
About the Author	8
Motivation	10
Introduction	14
Unification Theory	15
Structure of the matter or mass	17
Basic Building Blocks (B.B.Bs) of the Universe	18
Science and Religion	27
Weaving of Different Units of the Universe **(Fig. 4 to 7)**	33
Nomenclature of Atomic Genes [8]	34
Development of the Universe - Hyole-Narlikar Universe	36
Message System of the Universe	44
Message Network of the Universe (Feedback Mechanism and **Different Centers of the Universe)**	45

Cancer Cell is the Progeny of Normal Cell 48

Atomic Genetics and Phenomenon of Life Effects 49

How Does Nature Work & Triggering of Normal & Abnormal Life Effects 51

Triggering and Regulation of Cell Functions or Cell Physiology 52

Mechanics of Life and Death (see Fig - 23 & 24) 53

Cellular Oncogenes - I 58

Cellular Oncogenes - II 60

Atomic Genetics and Genetic Damage (see Fig-22) [9] 64

Atomic Genetic Engineering as Adjuvant Therapy & Message Formation in the Brain 76

Prayer and Growth Dynamics of Cancer 79

Predictions & New Observations of the New Theory 81

Comparative Study of Normal and Cancerous Mitotic Cycle (Fig-32) 83

Highest Center of the Universe 84

Glossary 86

References 89

Dedication

This book is dedicated to TWO BASIC BUILDING BLOCKS (which constitute First God of symmetry phase, without which universe could not be created -Mind and Mass part of the Truth) of the universe,. These are divine structural and functional units of the universe. These are eternal bodies of the Nature and they have transmutated themselves to form visible and as well as invisible Universe. Only I know and I worship these TWO OMNIPRESENT, OMNISCIENCE AND OMNIPOTENT.

Dr. Vijay Mohan Das

Discussing Mind and Mass with Prof APJ Abdul Kalam ex President of India and winner of Bharat Ratan - Sept -2013

Foreword

1. **Prof. S.W.Hawking**- His hopes for the attainment of a fundamental theory of nature, and its relevance to the general public, are best summed up in the concluding paragraph of his famous book: "... if we do discover a complete theory, it should in time be understandable in broad principle by everyone, not just a few scientists. Then we shall all, philosophers, scientists and just ordinary people, be able to take part in the discussion of the question of why it is that the universe and we exist. If we find the answer to that, it would be the ultimate triumph of human reason, for then we would know the mind of God".

2. **Prof. Roger Penrose**- In The Emperor's New Mind, a bold brilliant, groundbreaking work, he argues that we lack a fundamentally important insight into physics, without which we will never be able to comprehend the mind. More over he suggests, insight may be the same one that will be required before we can write a unified theory of everything.

Acknowledgement

I pay my gratitude to The Creator and Destroyer of the universe the yang B.B.B working as Highest center of the universe "I" by whose grace University of God came into existence. I am also thankful to Fritjof Capra who materialized Almighty B.B.B (His Avatar Form) in his Book Tao of Physics (figure 2) and also to Trinth Xuan Thuan and Collin A Ronan for materializing Almighty B.B.B. (His Avatar Form) in their books. Last but not the least my thanks to my teachers, my wife Dr Sapna Das, my family, friends and relatives and my son Dr Ankit Mohan Das, Dr Chandni my daughter, my daughter in law Dr Namita Nigam and my son in law Dr Abhishek for taking pains and having patience living with me and gave me moral support to complete this theory not only few years but also 27 years till its all volumes get published. We could not enjoy world rather they used to provoke me time to time to be aware of errors in completing this work. My wife Dr SapnaDas has great faith in God mainly Lord Shiva. She used to have fast during Navratri and have full devotion in Goddess Durga and her all nine avatars. My gratitude to my Father, late Shri Shyam Manohar Das and my Mother late Pushpa for being my parents and had full devotion in getting me brought up and making me surgeon.

About the Author

VIJAY MOHAN DAS was born at Fatehgarh, UP (India), on 18th July of 1957. He passed MBBS in 1980 with distinction in Physiology and M.S in 1984 from Bundelkhund University, Jhansi. He is a consultant surgeon and doing private practice in his hometown. He has done research work in the field of particle Physics, Astrophysics, Physiology of cell, and Oncology (Atomic Genetics And Basic Etiology of cancer). He has presented research papers in ATOMIC GENETICS AND BASIC ETIOLOGY OF CANCER in the following conferences.

1. ASICON-98, Dec.1998 at Ahmedabad
2. SGPGI-99, Feb.1999 at Lucknow. U.P.
3. Indian Society of Oncology, March.1999, at New Delhi.
4. UPASICON-99, October.1999 at Aligrah. U.P.
5. APCC-1999, December.1999 at Chennai (With International Particaption).
6. UPASICON-2000, NOV-2000 BANARAS, U.P.
7. WATCH-2000, DEC-2000 NEW DELHI (International Participation)
8. UPASICON-2001, to be held in NOV-2001, JHANSI,U.P. - To present the ORATION AWARD- Dr S.P. Shirivastava memorial oncology oration award - 2001.
9. 14th ASIA PASIFIC FEDREATION CONGRESS- INTERNATIONAL COLLEGES OF SURGEONS- 47th Annual Conference of Indian Section on 9th Nov. 2001 in HYDERABAD, INDIA to present paper on ATOMIC GENETICS AND BASIC ETIOLOGY OF CANCER as Dr. V.Mahadevan Memorial Best Research paper.

AWARDS

- 1. Dr. S. P. Shirivastava memorial oncology oration Award-2001 by ASI (Association of surgeons of India), U.P. chapter, Jhansi on ATOMIC GENETICS AND BASIC ETIOLOGY OF CANCER.

2. Published 10 volumes of titled, "Atomic genetics and Origin of the universe" in International journals printed and online in 2014.

3. You tube - link Science God Universe and Prayer - 15 videos

Dr.Vijay Mohan Das is Life Member of I.M.A., A.S.I. and I.P.A. (Indian Physics Association) and ISC (Indian Science Congress Association)

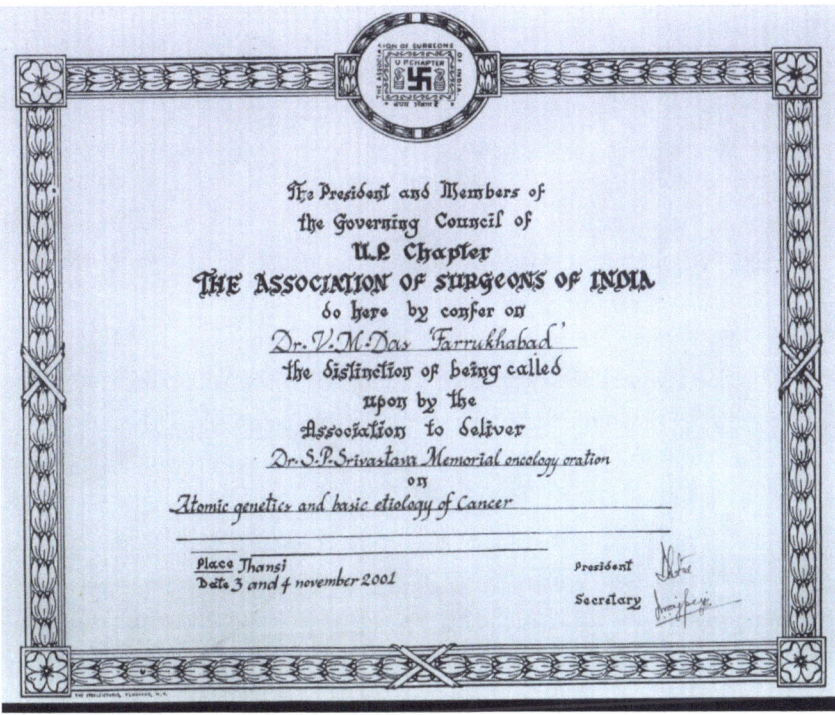

Motivation

Prof. J.V. Narlikar and Prof. Fred Hoyle had proposed continuous creation theory [1] in 1960s. The new model of the universe made by participatory science which is contrary to Big bang model supports theory proposed by Prof. Narlikar and Prof. Hoyle with addition of new scientific understanding calling it a **"NEW PHYSICS"**. Prof. Hermann Bondi, co-author of Steady State theory said, " It is 80% big bang, 5% steady state and 15% unknown ". New model of the universe not only explains all the events of Big bang and steady state but also 15% unknown events i.e. quasars and cold dark matter which is constituting 90% to 98% of the matter of the universe. Researches regarding structure of the matter are held up at lepto-quark level. Prof. J.V.Narlikar had asked to investigate the structure of the matter beyond leptons and quarks (the ultimate structure of the matter) in an article titled, **" Do Astronomical observation Require New Physics?"** in Physics News, Vol-30, N0-3&4, Sept &Dec, 1999 [2]. New Scientific understanding has explored the matter beyond lepto-quarks and ultimate structure of the matter i.e. Basic Building Blocks (B.B.Bs) are not only hypothesized but also there are observations that confirm their existence. Mathematics as well as Experimental Labs are required to know structure of the matter up to the level of leptoquarks. Beyond that it is the participatory science discipline which is required to know structure of the matter up to the level of Basic Building Blocks. As we have radioactivity, where nature is breaking itself to know about structure of the nucleus of the atom. Without this breaking it is not possible to study about nucleus. Similarly nature is breaking itself beyond leptoquark up to the level of Basic Building Blocks in the universe, only we have to re-explain those observations in terms of their constituents. These are-

- **Proton Decay** - As we observe decay of nucleus in radioactivity into alpha, beta and gamma, similarly proton does decay forming gravity and electromagnetic field particles. These field particles - energized gravitons (secondary fermions) are coming out from quarks while photons (secondary bosons) are coming out from decay of energized gravitons.

- **Gravity observations** - all gravity interactions should be re-explained. During these interactions energized gravitons interact by breaking themselves. By these observations we could know structure of the matter of secondary fermions and secondary bosons up to the level of primary fermions and primary boson.

- **Quasar observations** - Inside quasar Nature is breaking itself up to the level of Basic Building Blocks (CREATION PHYSICS). So we could see the basic constituents of all the force particles (except weak nuclear force which is mediated by vector bosons) We could see up to the level of Basic Building blocks (B.B.Bs)

- **Our Brain realization** - Inside our brain nature is working by breaking its last box i.e. atomic genes. Breaking of atomic genes, which is the property of the matter or basic building blocks, is triggering the thought process and other working of the brain. We could see up to the level of atomic genes property of Basic Building blocks.

Participatory science is a new discipline in science as proposed by Prof. John A. Wheeler [3]. Prof. John Wheeler sees this involvement of the observer as the most important feature of the quantum theory and he has therefore suggested replacing the word 'observer' by the word 'participator'. The idea of participation instead of observation has been formulated in modern physics only recently. Modern science teaches us up to the level of lepto-quarks. Beyond that, it is the participatory science that teaches us about structure of the matter upto the level of Basic Building Blocks (B.B.Bs) i.e. ultimate structure of the matter of which all fermions and bosons are composed. The entire participatory science has been developed by me and we could see structures which are beyond our visibility i.e. both macro (invisible universe) and micro (ultimate structure of the matter) worlds. Big bang and steady state models have been made because we could see only 30% of visible universe. Rest of visible and invisible universe could be seen through participatory science while making new model of the universe. Till today no attempt has been made to investigate about the consciousness of the matter. D. Bohm has found it is necessary to regard consciousness as an essential feature of the holomovement and it should be taken into account explicitly while considering this theory. He sees **mind and matter** as being interdependent and correlated but not causally connected [4]. Atomic genetics, a new concept in science has been introduced.

It is the study as regard mind part of the reality. There are observations that show that matter is related with consciousness. The most exciting observation is expansion of the universe shown in new model of the universe. Behavior of Dark matter is such that we are forced to assume that thought' is the inbuilt property of the cold dark matter to trigger expansion of the universe while making the new model of the universe. Before expansion or symmetry breaking phase, universe was in symmetry phase and to trigger symmetry breaking, it is the 'thought' an inbuilt property of the matter or B.B.Bs (of entire universe), which is responsible for this triggering too. The problem of how matter attained masses has been meticulously solved by the research of B.B.Bs. Matter attained mass by virtue of mass property of B.B.Bs. So the God's particles are Basic Building Blocks (mind and mass unit) rather than Higgs bosons as proposed by Prof. Peter Higgs in STANDARD MODEL.

Discussion and Inferences

Science has not yet defined God. Co-relation of science and religion can be made possible after the concept of Basic Building Blocks is well understood. Religion guides us in recognizing these B.B.Bs. The model of B.B.Bs is made on the basis of inertial properties of energy and matter. On the same fundamental basis religion had incorporated certain definite clues thousands of year back. When a parallel is drawn between the two (models made by participatory science and the clues given by the spiritualism) by using common logic, incorporation of science and spiritualism can thus be made possible (Fig 13). So far no attempt has been made to define eternal properties of energy and matter at the level of B.B.Bs. No attempt has been done to investigate **PURE matter**. Fermions are **IMPURE matter** as they have spin properties. The research of B.B.Bs or mind and mass would produce fragrance of God (B.B.Bs or Omnipresent) in the new model of the universe. Thus Einstein's question that how God created the universe can be meticulously solved by introduction of the new model of the universe. New model of the universe shows that the universe is deterministic universe and all quantum, classical and life sciences effects are triggered by thought expressions or atomic transcriptions **(cause and effect concept)** and thus their precise prediction by the participator (B.B.B working as highest center of the universe) could be possible in future. Thus Einstein's famous metaphor that **God (B.B.B) does not play dice** ultimately became the truth along with final acceptance to **Laplace determinism—** All that had happened had a definite cause and gave rise to definite effect and future of any part of the system could in principle be predicted with absolute certainty if its state at any time was known in all details [5].

The final stamp of success to the new model of the universe has been given by observation published in journal JAMA [6] by the study- prayer helps cardiac patients. Prayer is now a well-confirmed phenomenon and it is related with God (B.B.Bs working as highest center of the universe). It is the replicated study. Phenomenon of prayer profunds that hypothesis of new model of the universe in which one B.B.B is working as highest center of the universe is correct and phenomenon of feed back to this B.B.B exits in this universe. This phenomenon of prayer also propounds that there was a precreation era in which programming of future universe was done by highest center (B.B.B) of the universe. If the phenomenon is replicated, the supporting theory is believed to be the truth. Big bang and its early events (GUT, and super unification) that prove this theory neither could be replicated in lab nor could be observed (replicated observations) anywhere in the universe. On the other hand, continuous creation as proposed by Prof. Narlikar and Prof. Hoyle could be observed (replicating phenomenon) in quasars and also it is replicating every time. New model of the universe based on different observations not only supports this idea but also it could prove how creation is going inside quasars.

Hoyle and Narlikar proposed (in their continuous creation theory) that new matter is being created due to 'IMPLOSION' to balance the expansion of the universe, which astronomers have observed. Inside 'QUASI-STARS' gravitational collapses may form some matter in the universe. The huge luminosity and the radio emission from these quasi-stars appear to be 'gravity powered' unlike ordinary stars, which derive their energy from nuclear reactions.

Einstein preferred to believe that the universe was ageless and eternal [7]. Einstein's views were correct when they are applicable to symmetry phase of the universe. After symmetry breaking phase only small part of entire universe got into expansion phase along with creation of the matter, both hot and cold. After the contraction phase of the universe, it would again go into symmetry phase and then entire universe would be not only ageless and eternal but also it would be infinite, absolute and **holomovement** –to which participatory science calls **"I" or One Absolute "I" made up of two God Particles.**

Introduction

To know basic etiology of cancer, one must know the structure of the matter or mass, origin of the universe and atomic genetics as taught by newly developed science called **PARTICIPATORY** science. The whole research is presented in thesis form involving experiments, observations and inferences. The outcome of the research is discussed in very simple way just to acquaint reader about recent advances in particle physics, astrophysics, physiology of cell, oncology and finally in the treatment of the cancer. Unless one answers following question, one can not know basic etiology of cancer.

'I WANT TO KNOW HOW GOD CREATED THE UNIVERSE
I'M NOT INTERESTED IN THIS OR THAT PHENOMENON..
I WANT TO KNOW HIS THOUGHTS.
THE REST ARE DETAILS.'

-ALBERT EINSTEIN

Unification Theory

The highest hypothesis in physics is unification theory **Fig 1**. It is also called Einstein's dream. The theory has predicted super unification phenomenon before origin of the universe. It does mean that four natural forces which are mediated by four different particles (gravitons, gluons, vector bosons and photons) were identical or belonging to the same family. It is also called **ONE PARTICLE THEORY**. This theory is not proven yet. One particle theory means that are natural forces are created from one type of particles.

Now I am proposing **TWO PARTICLES HYPOTHESIS**. It does mean that four natural forces which are mediated by four different particles are created from two type of particles instead of one types. So finally, there are two views. One is Einstein's dream i.e. one particle theory and the other one is participatory science's view i.e. two particle theory

Science, God, Universe and Prayer

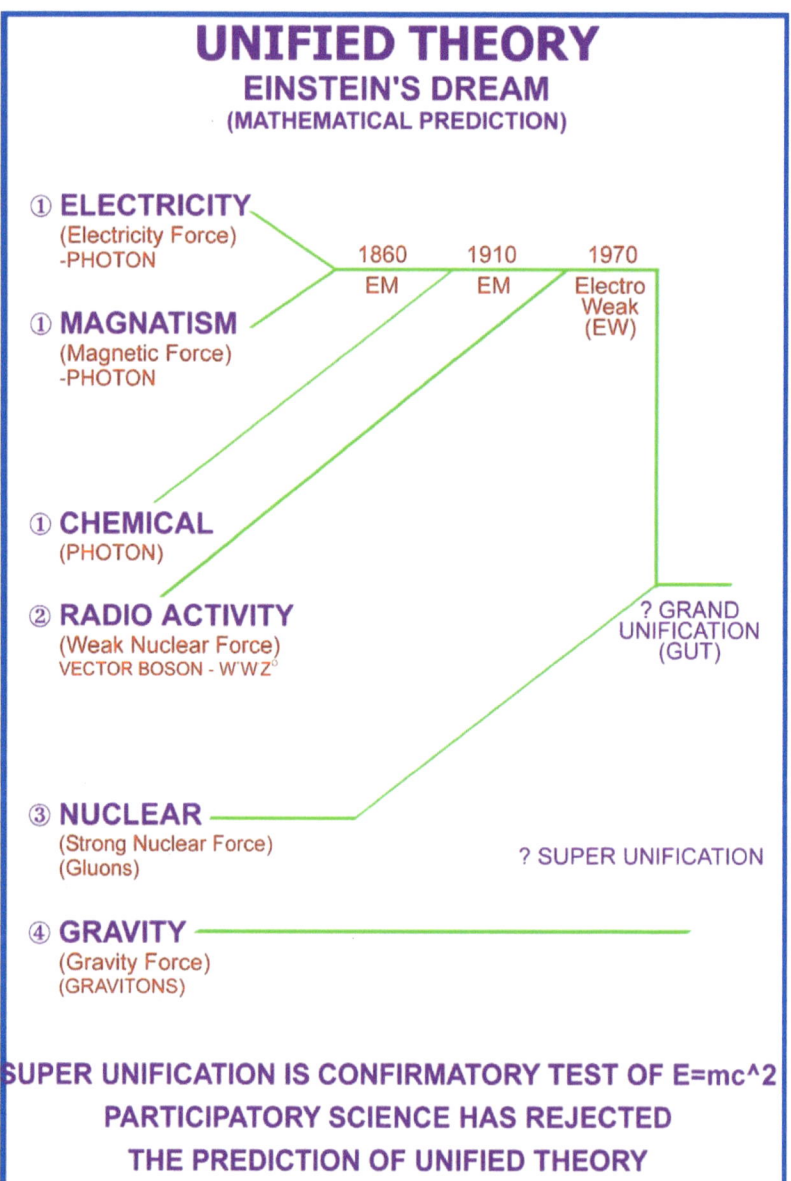

Fig-1

Structure of the Matter or Mass

Matter is made up of molecules, molecules are made up of atoms, and the atoms are made up of nucleus and the electrons which revolve around the nucleus. Electrons are also called leptons. Nucleus is made up of protons and neutrons and protons and neutrons are made up of quarks. So lepto-quarks are supported to be the smallest matter particles. But according to participatory science, lepto-quarks are not fundamental particles of the matter.

See Figure. 2

We have explored the nature beyond lepto-quarks. We have found that lepto-quarks are made up of energized gravitons and secondary bosons. Energized gravitons are made up of gravitons and primary bosons. Finally, gravitons are made up of two types of basic building blocks (B.B.Bs) called matter B.B.B. and energy B.B.B.. While primary bosons are made up of one type B.B.Bs. called energy B.B.B.. The other name of these B.B.Bs. are YANG (matter B.B.B) and YIN (energy B.B.B.).

See Figure.3

The size of the mass of these particles is reducing and then B.B.Bs. are the smallest mass particles of which all fermions and bosons are composed. From these fermions and bosons all the matter (visible as well as invisible) of the universe is formed including the human cells. Though the Figs.4 to Fig.8 are self explanatory some details will be given later.

See Figure.4   See Figure.5   See Figure.6
See Figure.7   See Figure.8

Basic Building Blocks (B.B.Bs) of the Universe

These B.B.Bs. are the smallest structural and functional units of the universe. Upon these B.B.Bs., ATOMIC GENES are found. These basics units are divine in the sense they talk with each other by phenomenon called atomic transcription and translation. These are fundamental particles and atomic transcription and translation is fundamental working of the nature. These B.B.Bs have power to transmutate to form any bigger unit of the universe like particle, atom, molecule, complex molecules of the life, organalle, cell, tissue, organ, system, individuals, earth, solar system, galaxy, etc.,etc.. So all effects of the universe are triggered by atomic transcriptions or thought expressions.

The properties of these two basic units are opposite. This would be explained during the discussion of inertial properties of the matter while giving definition of the energy (E) and matter (m) of the expression $E=mC^2$ Therefore, this is called unity of opposite or unity of complementary particles.

See Fig.9

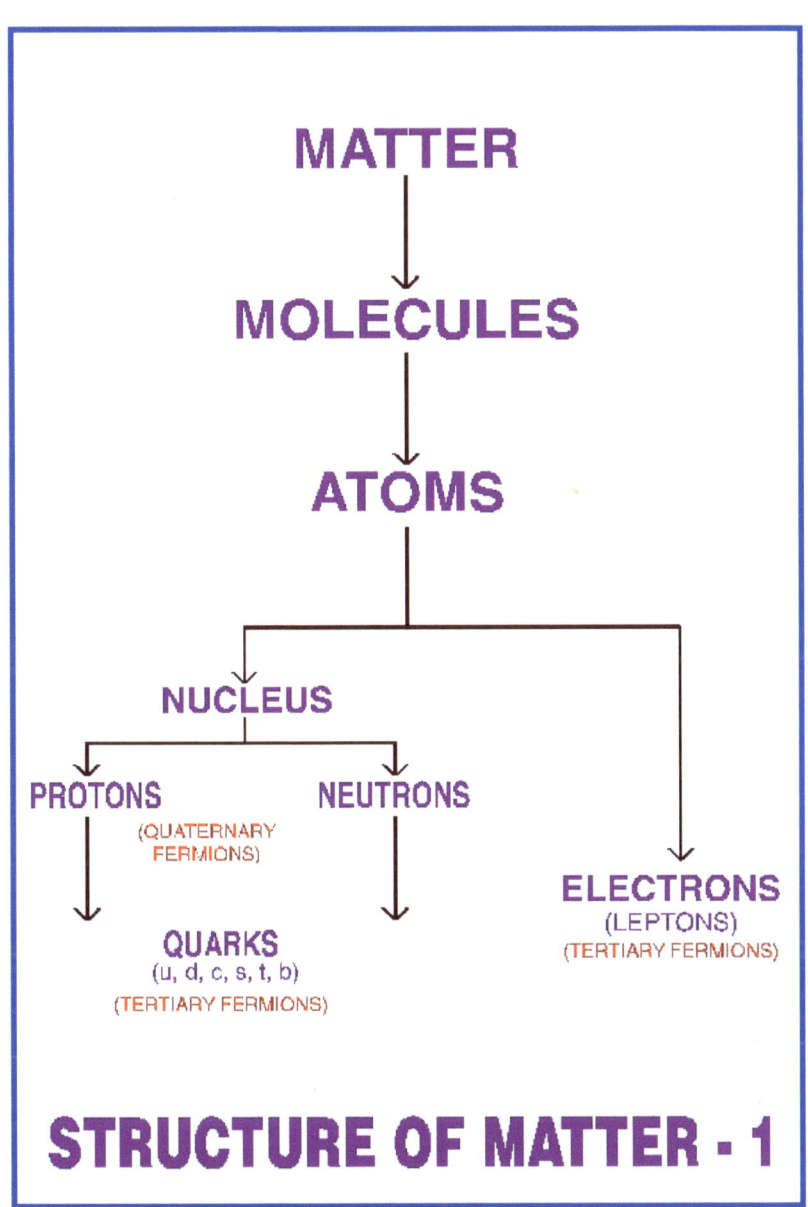

Fig-2

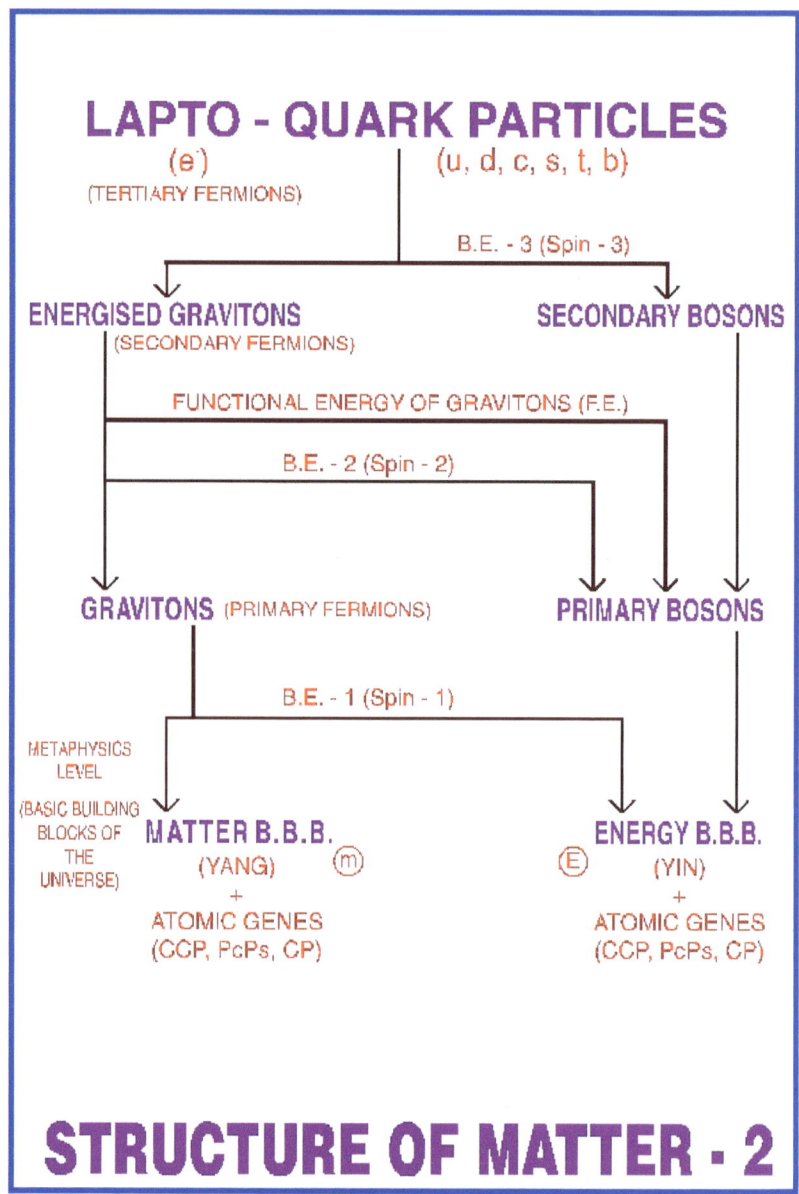

Fig-3

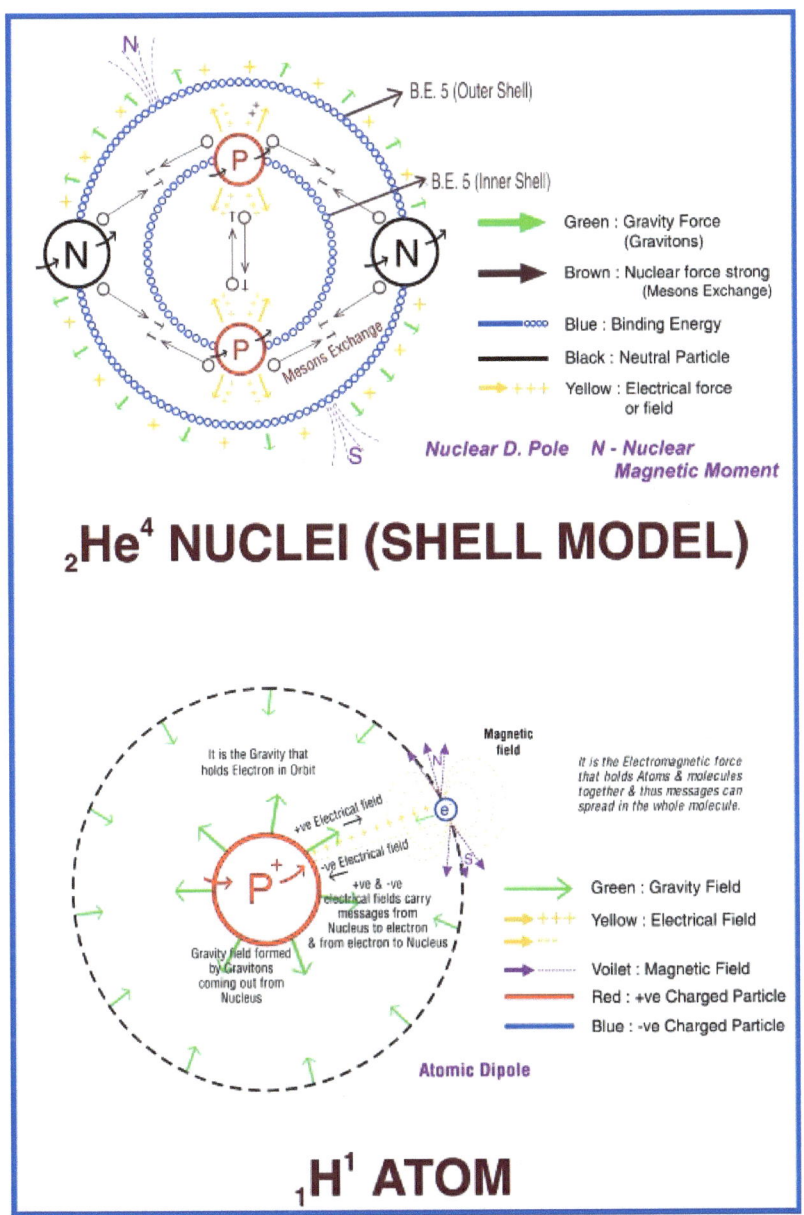

Fig-4

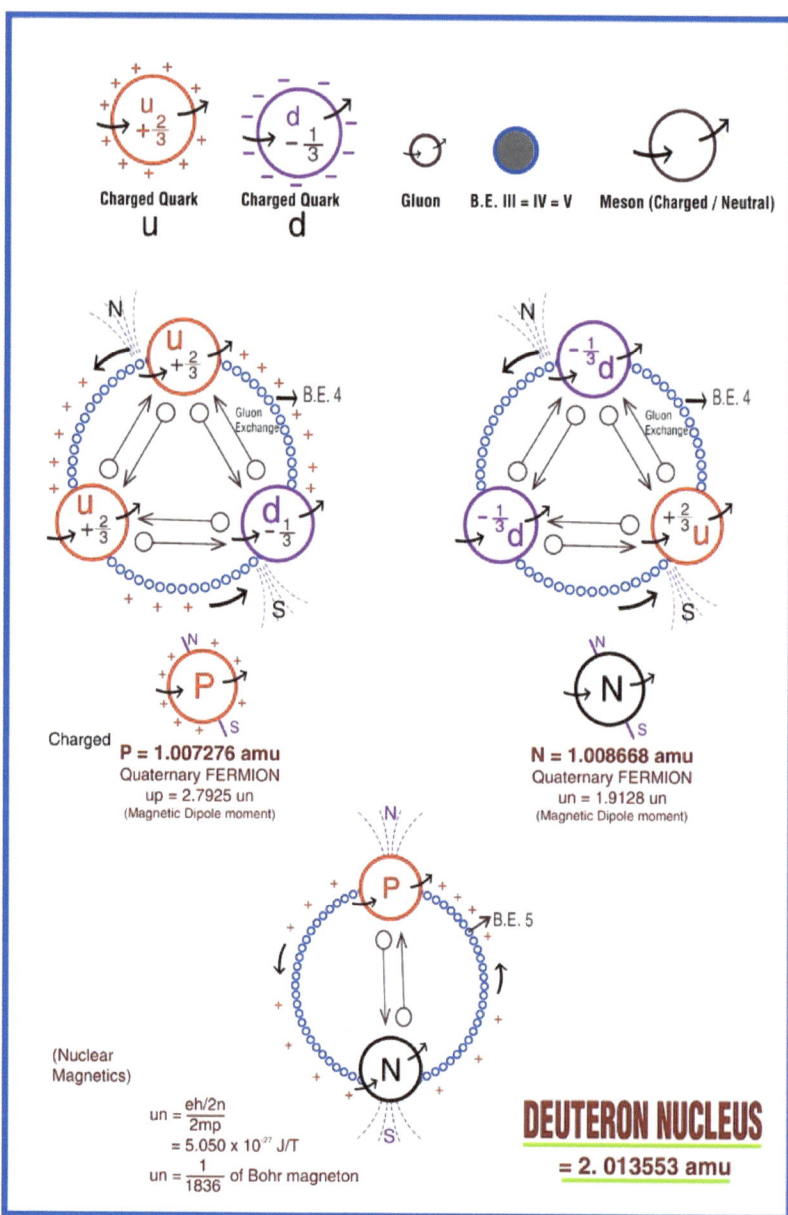

Fig-5

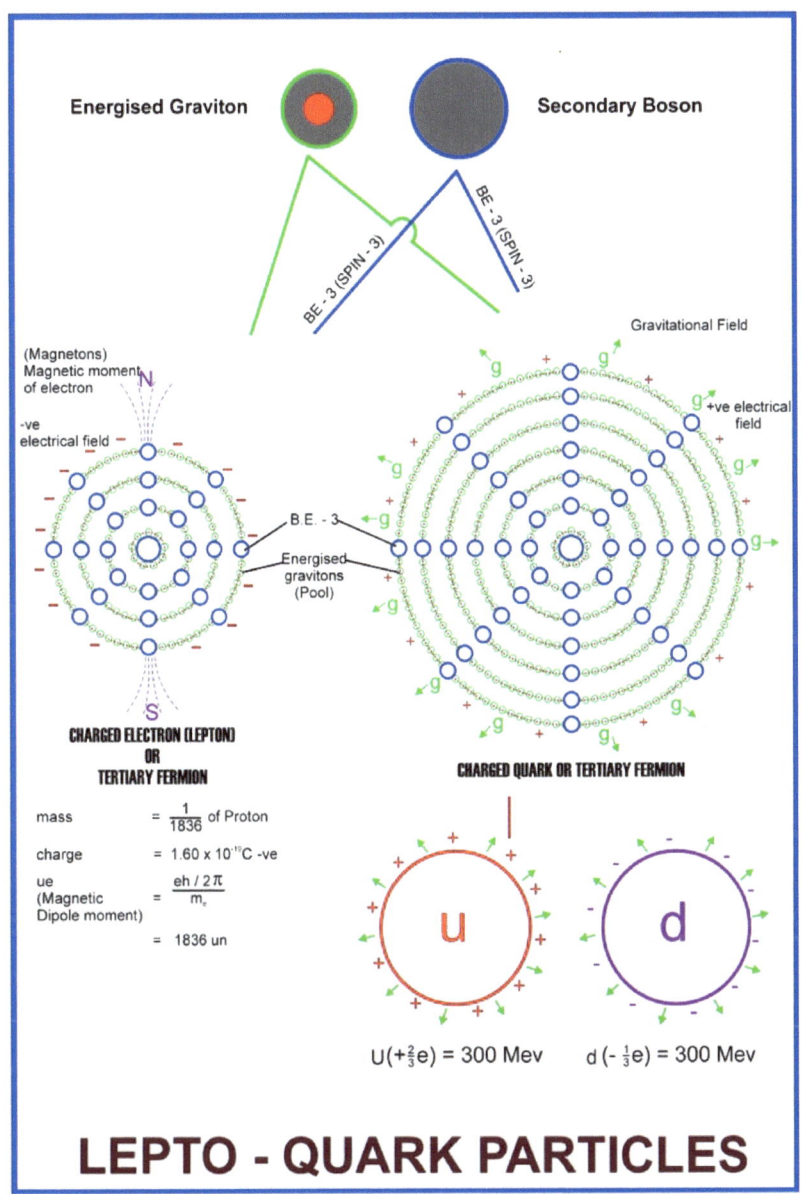

Fig-6

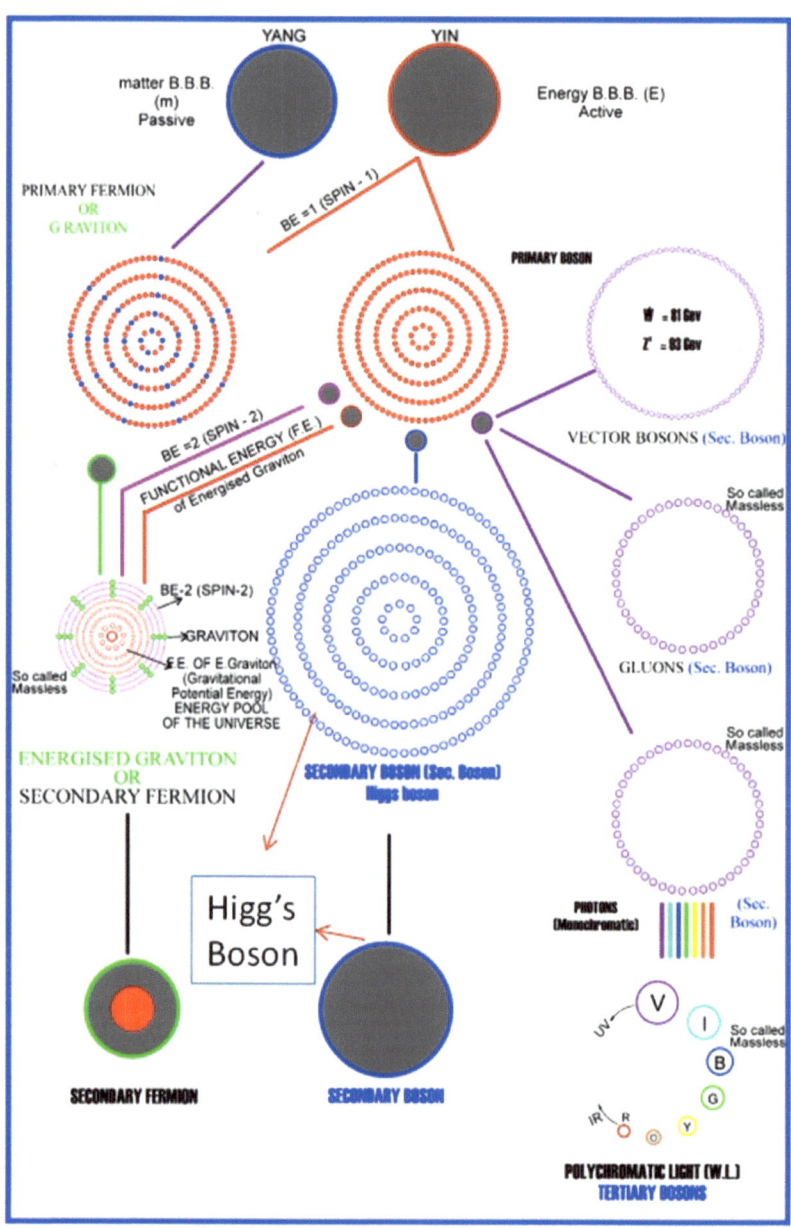

Fig-7

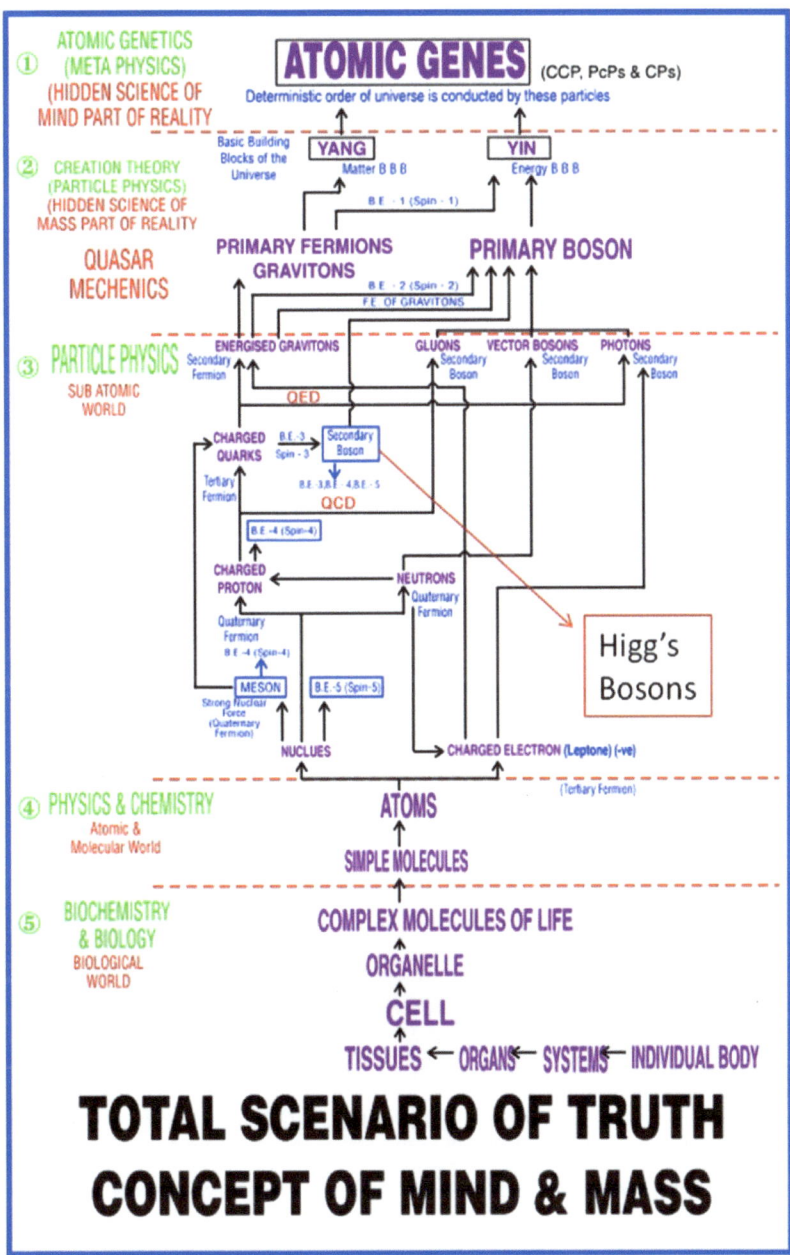

Fig-8

MODEL OF BASIC BUILDING BLOCKS OF THE UNIVERSE

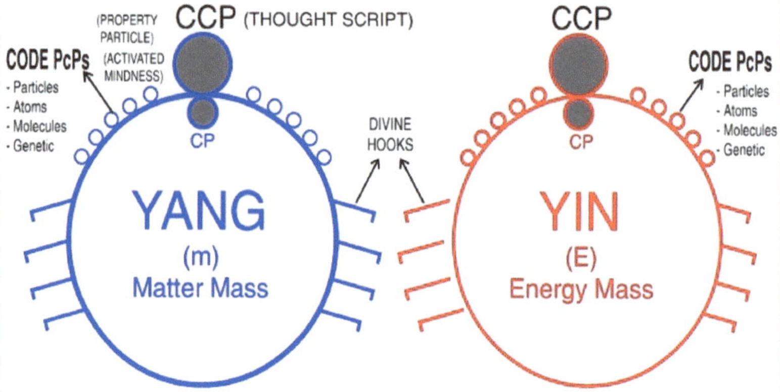

(PROPERTY PARTICLE)
CCP (THOUGHT SCRIPT)
CCP

CODE PcPs (ACTIVATED MINDNESS)
- Particles
- Atoms
- Molecules
- Genetic

CP

DIVINE HOOKS

CODE PcPs
- Particles
- Atoms
- Molecules
- Genetic

CP

YANG
(m)
Matter Mass

YIN
(E)
Energy Mass

CP is shown in the Picture
(Realisation Particle)

UNITY OF OPPOSITE

BOTH ARE COMPLEMENTARY

DIVINE B.B.B.s

MIND & MASS REALITY

(ATOMIC GENES)

(MATTER MASS ENERGY MASS)

Fig-9

Science and Religion

Spiritualism guides us in recognizing these two B.B.Bs or fundamental units. Hindu spiritualism has given us two clues which are revealed through some old paintings. These clues are based on basic properties of B.B.Bs. How these paintings or pictures evolved is a matter of separate discussion. But at present I would compare them through their representative colors. Let us call these paintings as clue pictures.

Modern physicists have not defined energy (E) and matter (m) in terms of their eternal properties. On this fundamental basis Hindu spiritualism has made these clue picture. Thus incorporation of the science and religion is made possible. Modern physicists have explored the nature of some particles (Fermions and bosons) only. They teach only properties of fermions (impure matter) and bosons (energy). There is no effort to know the properties of the B.B.Bs.

The first clue picture is androgynous from of "I". It is the clue as regard to the mass part of the reality. It does mean that there are two types of masses existing in the universe one is matter (m) mass and other one is energy (E) mass. The properties of the two are opposite. That is why one is shown as male and the other one is shown as female (unity of opposite). The blue B.B.B. is represented as blue body while red B.B.B. is represented as red body of androgynous form of "I".
See Figure.10

The second clue picture is God Form of "I" as shown in GEETA (Fig.11/12).It is the clue as regard to the mind part of reality. It does mean that mind or conscious particles or atomic genes are found in the nature. As this picture does not show two particles hypothesis, therefore we have modified it and replacing the central head with side head of Shiva. Further the body of Lord Krishana is replaced by androgynous form of "I". The final truth has emerged like this as shown in the picture of final truth. It is called Mind and Mass part of truth.
See Figure.13   see Figure.11   see Figure.12

The Final Truth: 'Mind & Mass Part of Reality'

Models of B.B.Bs. made by participatory science and compared with the final truth picture given by the Hindu spiritualism. Using our common sense we find that blue body of matter B.B.B. is the blue body of male and red body of energy B.B.B. is the red body of female. The central head of male is nothing but CCP (cosmic conscious particles) atomic gene of Yang (matter B.B.B.), central head of female is nothing but CCP atomic gene of Yin (energy B.B.B). The side heads of final truth are nothing but code PCPs (programmed conscious particles) atomic genes of Yang and Yin B.B.Bs. The black bodies four on each side which are there in final truth, are nothing but CP (conscious particles) atomic gene of B.B.Bs. The multiple hands of final truth are nothing but divine hooks of the B.B.Bs... So if we observe both the models using our common sense, the models of the B.B.Bs. made by participatory science are nothing but Final Truth picture given by the spiritualism.

See Figure.9   See figure 13

Clue Regarding Mass Part of Reality

**Concept of YIN (Energy Mass) & YANG (Matter Mass)
Unity of Opposite**

Fig-10

Clues Regarding Mind (Atomic Genes)
Part of Reality

Fig-11

Fig-12

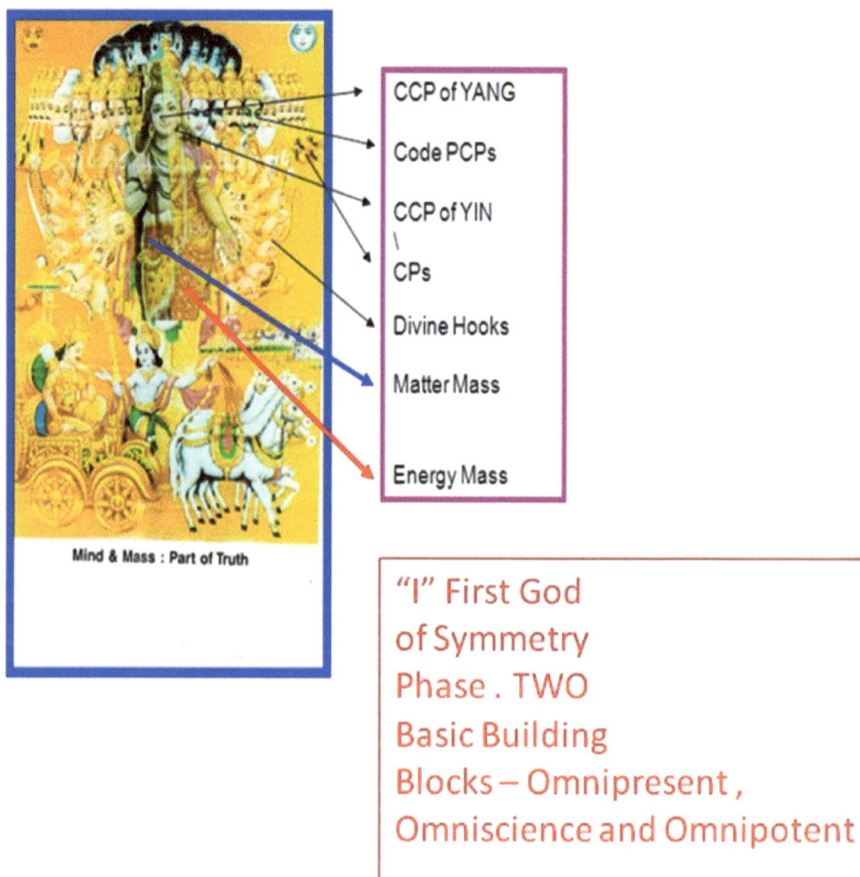

Fig-13

Weaving of Different Units of the Universe
(Fig. 4 to 7)

It is these divine basic units that have woven all the structures of the universe. These two B.B.Bs. have joined to form gravitons, while energy B.B.Bs. have joined to form primary bosons only. Later, gravitons and primary bosons have joined to form energized gravitons, while primary bosons only have joined to form secondary bosons (photons, vector bosons, gluons). Binding energy (B.E) (Higg's bosons) is nothing but a type of energy which is responsible for binding the matter B.B.Bs to one another along with marking the created particle (fermions) to spin. The details would be given during the discussion of creation physics. Later energized gravitons and secondary bosons have joined to form lepto-quarks. Quarks along with secondary bosons from protons and neutrons. Finally electrons revolved around proton forming atom.

It is these divine B.B.Bs. which are transmulated into atoms. Atoms have joined to form molecules, complex molecules of life, organelle, cells, tissues, organs, systems, different individuals. Similarly, atoms have transmulated into planets, solar systems, galaxies, super galaxies and invisible universe. So, basically these divine B.B.Bs. are every where in the universe or we may say they are OMNIPRESENT. The omnipresent is defined as GOD in all religious books. For example chapter 10/8 of GEETA-

I AM THE ORIGIN OF ALL, EVERY THING EVOLVES FROM ME KNOWING THIS THE WISE WORSHIP ME.

So. basically these B.B.Bs. are nothing but l of GEETA as every bigger unit of the universe has been originated from these two B.B.Bs. See Figure.13

Nomenclature of Atomic Genes [8]

In participatory science, we take similarities from biological world to understand the B.B.Block world. One who knows biological transcription and translation, can also understand atomic transcription and translation.

- Cell is anatomically and physiologically unit of the body. Similarly B.B.Bs. are structural and functional units of the universe.

- Cell functions (metabolic) are controlled by biological transcription and translation. Similarly, B.B.Bs. functions are controlled by atomic transcription and translation.

Biomolecules take part in biological transcription and translation are
See Figure.14

- DNA- message storage system

- mRNA-messenger molecule (carries message from nucleus to cytoplasm)

- Ribosome- translating molecule (it translates the message and works accordingly).Similarly, atomic genes that take part in atomic transcription and translation are-
  See Figure.9   See Figure.13

  - CCP-thought storage system (omniscient). It is similar to DNA of the biological world

  - Code PCPs-messenger atomic genes. It similar to mRNA.

  - CP-translating atomic genes. It translates the messages and realizes the message and react accordingly.

Fig-14 [8]

Development of the Universe - Hyole-Narlikar Universe

Before the origin of the universe, these B.B.Bs. were in the form of tachyons. It does mean that the tachyons were every where in the universe. Let us look at the structure of tachyon, it is made up of one of matter B.B.B. (YANG) and many energy (YINs). Initially out of the infinite tachyons, one became the highest center of the universe. Messages used to go from highest center to rest of the universe and messages could come from rest of the universe to highest center of the universe by atomic transcription. Thus highest center had fed its thought to rest of the B.B.Bs. that would take part in creation - that they would express only those thoughts to give desired effect as wished by the highest center of the universe. So all B.B.Bs were informed about their role before creation of the universe. In pre-creation era programming of the future universe was done by highest center of the universe.

See Figure.15(i)   See Figure16

Our universe is oscillating and it is a divine universe. It means that it has a creation phase and a destruction phase. During creation phase tachyons break into their B.B.Bs. and from these B.B.Bs, formation of fermions and bosons take place. After the creation phase, destruction would start and in this phase all created particles would again break into their B.B.Bs. and finally tachyons would form.

See Figure.16

At the time of origin of the universe, the effects got created. These effects are taking of different shapes and appearance of properties and law. Both these effects are studied in different branches of science. The cause of all the effects is THOUGHT. It does mean that unless the thought is expressed, programming is done, the messages reach to target B.B.Bs., nature cannot take new shape, properties and laws. Change in shape, properties and laws is called TRANSMUTATION. During transmutation target B.B.Bs. show synchronized working of mind and mass. With the result nature takes new shape, properties and laws.

See Figure.17

Thought (atomic transcription), programming (formation of messages), interaction (spread of massage) take place on B.B.Bs.. Therefore it is not visible to us. Only the effect part is visible or observed. In each and every transmutation, thought expression (atomic transcription) is the essential step. Thought, programming and interaction are collectively called as CCP.So in each step of transmutation, CCP is written. It does mean that unless the thought is expressed, nature cannot be transmutated.

See Figure.17

Nature takes shape due to mass property of the B.B.Bs.. While appearance of properties and laws are due to different types of thought statements or atomic transcription and which is due to atomic genes. The details shall be presented in chapter of ATOMIC GENETICS

See Figure.17

With the origin of the universe, nature first created a sphere of COLD DARK MATTER (C.D.M) and canals in it. With the result space got created. At the other end of the canals, hot reaction started. As a result hydrogen clouds and lot of radiation were created. The empty canals were filled by these hydrogen clouds and radiations and thus QUASARS appeared in the universe. Simultaneously C.D.M. layer started expanding and clouds and radiations kept on coming in this closed universe. With the passage of time more and more C.D.M. layer formed, more and more quasars formed. The hydrogen cloud came in this closed universe. They started running towards C.D.M. layer as they were attracted by the gravity of C.D.M. layer. The HUBBLE LAW can thus be explained

See Figure.15(ii,iii & iv)   See Figure.18   See Figure.19

With some more passage of time, clouds were joined to form GMC (giant molecular clouds). Later by self gravitation different proto stars, proto planets, proto satellites were formed. Finally stars became bright and thus bright galaxies appeared in this universe. Our universe is still in expansion phase and creation is still going inside quasars. It is to be remembered that highest center of the universe does not come in the visible universe. It keeps on receiving the messages by atomic transcription and it has power to change any programming programmed by it during pre-creation era.

See Figure.15(iv).

CONTINUOUS CREATION THEORY—- HOYLE AND NARLIKAR THINK THAT NEW MATTER IS BEING CREATED DUE TO ' IMPLOSION' TO BALANCE THE EXPANSION OF THE UNIVERSE, WHICH ASTRONOMERS HAVE OBSERVED.INSIDE 'QUASI STRARS' GRAVITATIONAL COLLAPSES MAY FORM SOME MATTER IN THE UNIVERSE. THE HUGE LUMINOSITY AND THE RADIOEMISSION FROM THESE QUASI -STRARS APPEAR TO BE,GRAVITY POWERED' UNLIKEORDINARY STARS WHICH DERIVE THEIR ENERGY FROM NUCLEAR REACTIONS.

Prof. Fred Hoyle (1915-2001)   Prof. J.V.Narlikar (Director, IUCAA, Pune)

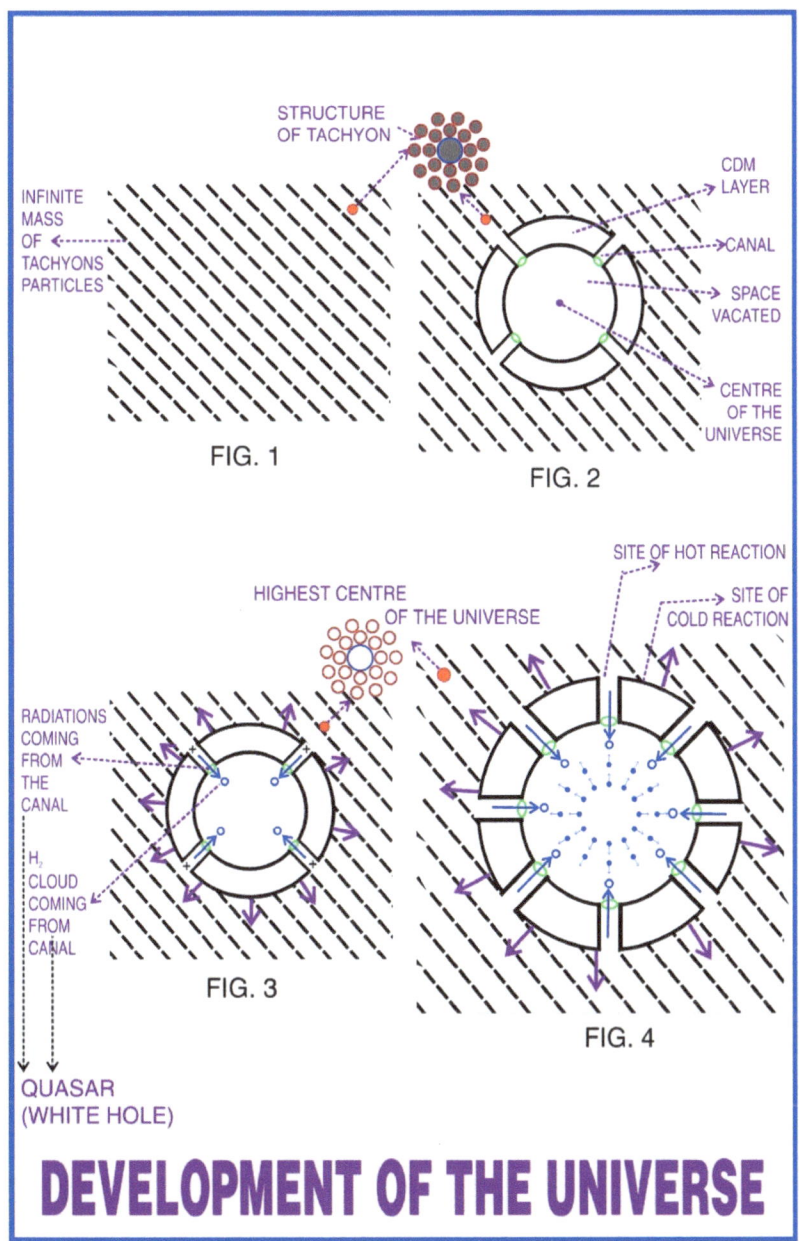

Fig-15

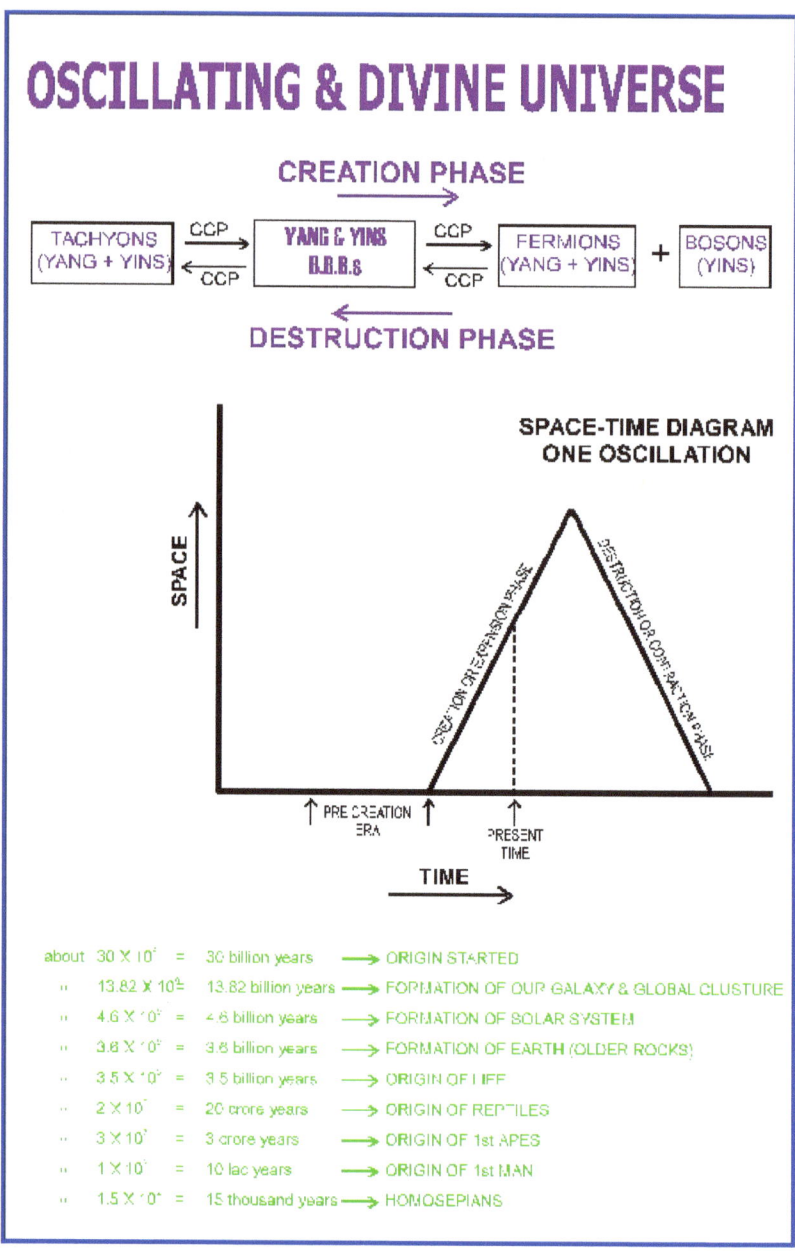

Fig-16

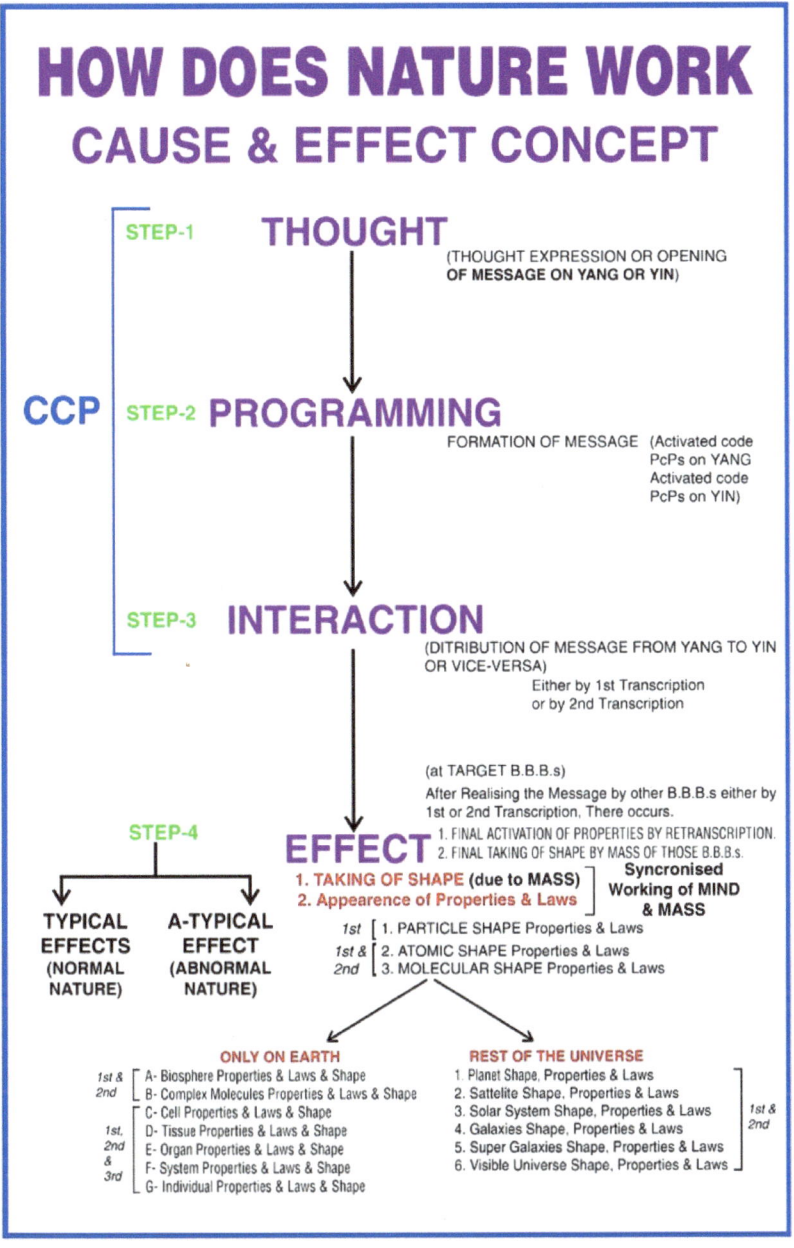

Fig-17

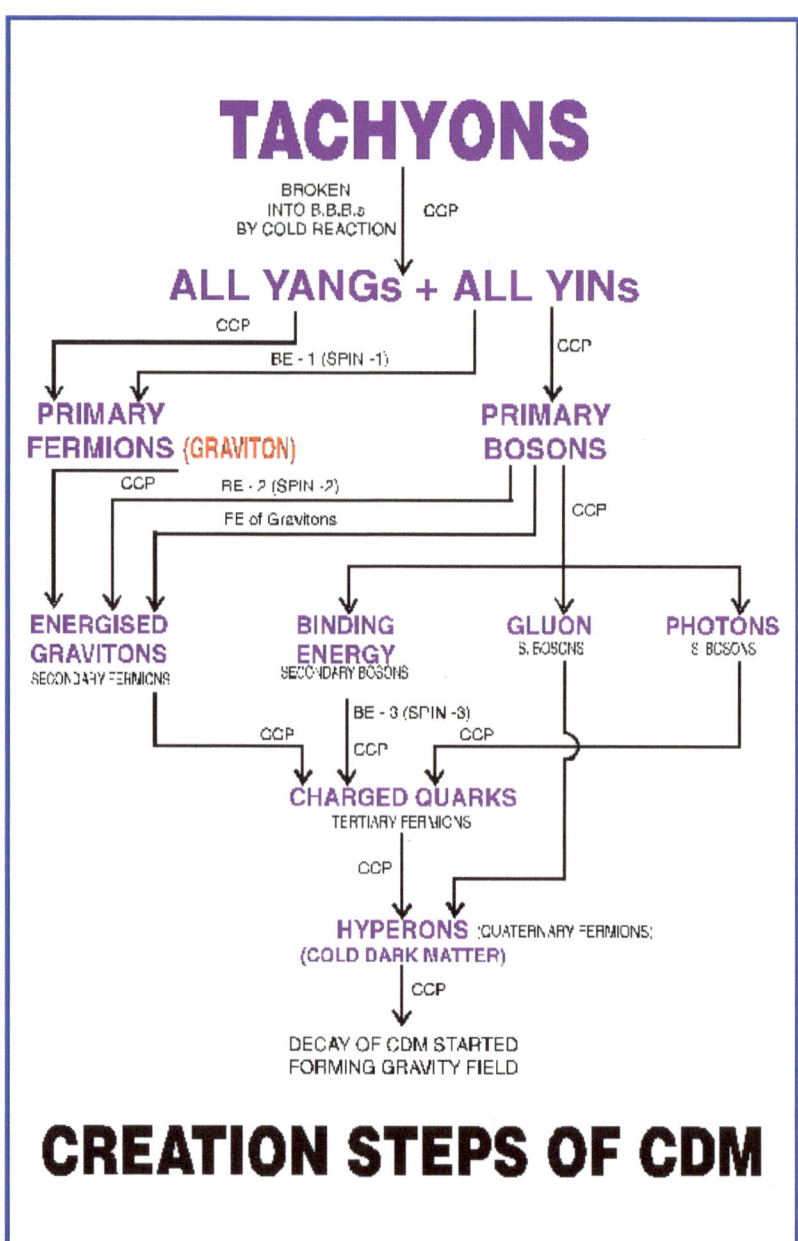

Fig-18

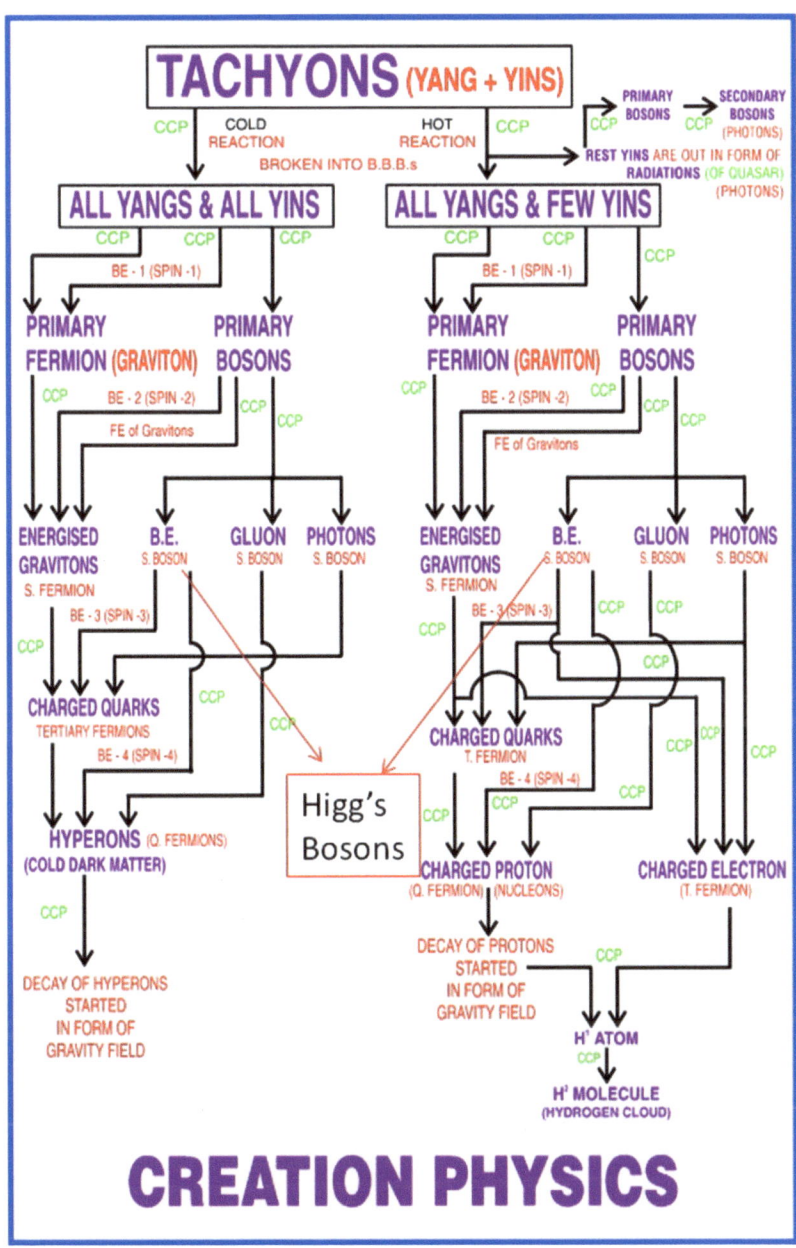

Fig-19

Message System of the Universe

Before the origin of the universe nature had only one type of message systems which is called FIRST TRANSCRIPTION. Messages used to go from one B.B.B. to another B.B.B. by atomic transcription. Messages were carried by atomic genes with very very high velocity. It is the fundamental message system.

After the origin of the universe, nature created atoms. It also created one more message system called SECOND TRANSCRIPTION. Here the message (code Pcps) are carried by photons from one atom to another atom with velocity of light. Thus atoms,molecules, cells, and even individuals talk with one another After the formation of the cell, nature created one more system called THIRD TRANSCRIPTION.Here there is a message storage system formed by DNA. There are messenger molecules called mRNA that carry message from DNA script to cytoplasm where the message (code PCPs) is read or translated by ribosome and they work accordingly. Thus the messages reach to enzymes and hormones and finally messages reach to target units. Having received the messages, target units work accordingly. Finally life effects (metabolic) are observed.

These three types of message systems which are working in the nature. These message systems are being used by the nature according to nature's need.
See Figure.20

Message Network of the Universe (Feedback Mechanism and Different Centers of the Universe)

With the origin of universe, nature first created primary units i.e. primary fermions (gravitation) and primary boson, these primary units are equipped with one higher center (one B.B.B.) and rest of the B.B.Bs. are working as lower centers or target B.B.Bs. After primary units,nature created secondary units i.e. secondary fermions and secondary bosons. similarly nature created tertiary units (lepto-quarks) and then quaternary units (protons & neutrons). Each unit is equipped with higher centers, lower centers and target B.B.Bs. After quaternary units nature created atomic units, molecular units, complex molecules of life units, organelle units, cell units, tissue units, organ units, system units and individual units. Each unit is equipped with higher centers, lower centers, and target B.B.Bs. Similarly nature created satellite units, planet units, solar system units, galaxy units, super galaxy units, dark matter layer unit. These units are also equipped with higher centers, lower centers and target B.B.Bs. Thus our universe is divided into different units and each unit is equipped with higher and lower centers.

All higher centers are under control of highest center of the universe by efferent paths. This efferent path is made up of first transcription. Higher centers can send messages to highest center of the universe by afferent path or feed back path. Thus highest center of the universe is well informed about all effects of the universe. Messages can come from lower centers to higher centers and from higher centers to highest center of the universe via afferent path. The highest center of the universe can send messages to higher centers and from higher center to lower centers. There is an inter unit message network also which is made up of first, second and third transcription depending upon the nature's need. Thus the entire universe is under control of highest center of the universe. Highest center can change any programming programmed by it during pre

creation era. See Figure.21

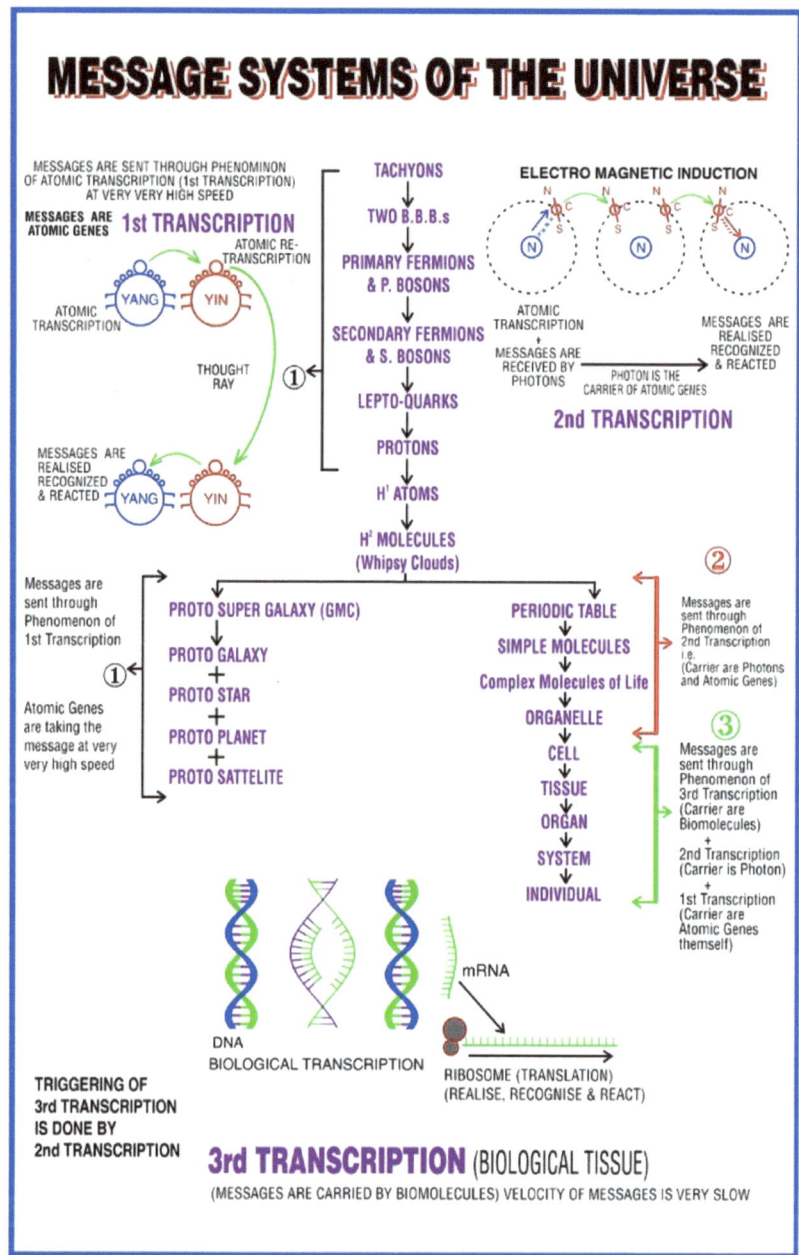

Fig-20

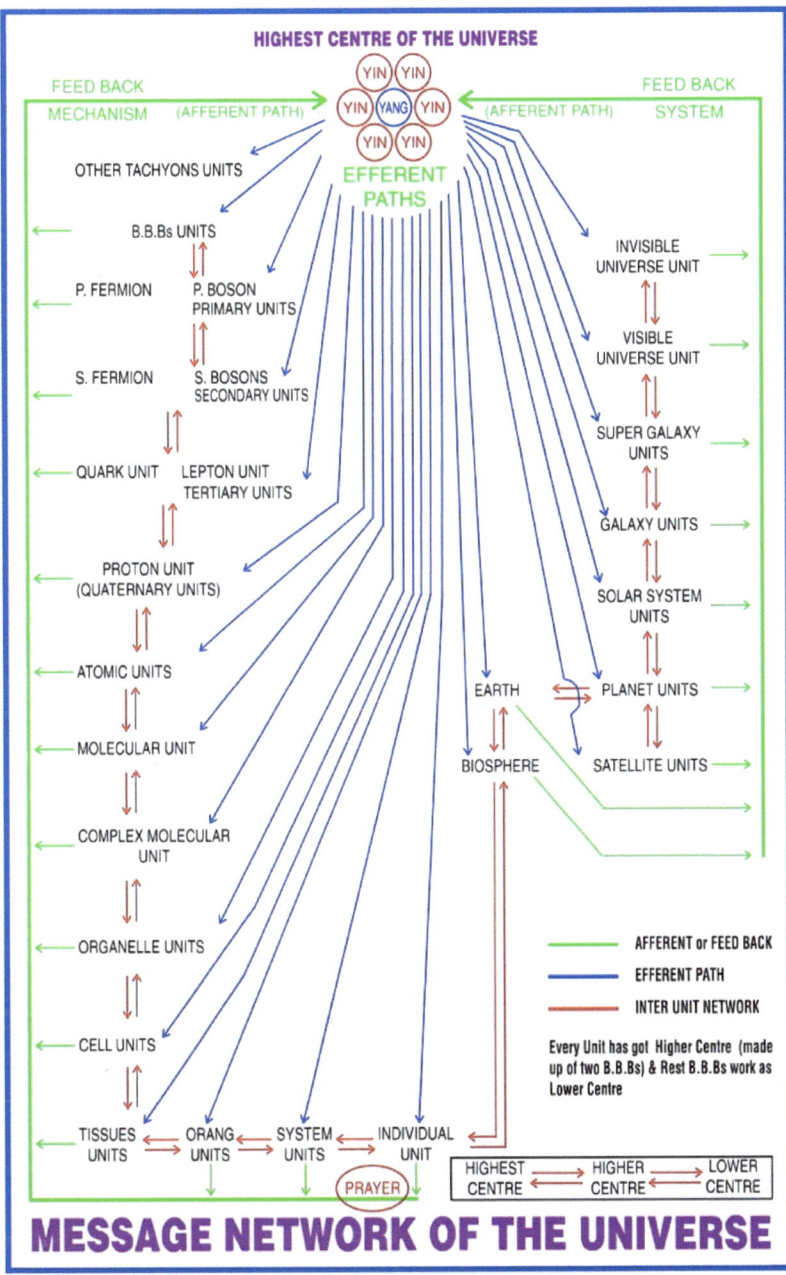

Fig-21

Cancer Cell is the Progeny of Normal Cell

The normal cell structure and functions are triggered by normal thought expressions. With the result there are programmed messages of normal genetic configurations, normal growth and differentiation. These programmed messages which are carried by code PCPs are received by target B.B.Bs. of the cell. They start showing normal genetic configurations, normal growth and well differentiation normal cell structure. During transmutation, these thought expressions are suppressed and abnormal thought expressions are triggered with the result there is formation of programmed messages of genetic damage, rapid growth and de-differentiation. These abnormal programmed messages which are carried by code PCPs are received by same target B.B.Bs. which are previously showing normal effects. The same B.B.Bs. now start showing abnormal effects like genetic damage, rapid growth and de-differentiation. With the result normal cell transmutate into cancer cell. There are millions of normal cell effects but for understanding the concept I have taken only three effects. Similarly, there are millions of cancer cell effects that are triggered by millions of abnormal thoughts expressions. But for understanding the concept I have depicted only three abnormal effects.
See Figure 22

Atomic Genetics and Phenomenon of Life Effects

Being a biologist, one must know how do life effects come about. Life effects are higher thought expressions of B.B.Bs. Formation of particles, atoms and molecules is due to lower thought expressions of B.B.Bs. But their higher thought expression lead to appearance of all life effects. One who knows properties of B.B.Bs. and atomic genetics, can understand how life effects are triggered. There is nothing like SOUL. It is a myth that when soul goes inside we get life and when it moves out we are dead. When thought expressions of life are suppressed and thought expressions, of death are triggered, we observe death effects. So life effects are basically triggered by atomic transcription occurring on B.B.Bs. If life effects are normal leading to normal cell structure and functions, it is due to normal thought expressions. And when thought expressions are abnormal, it leads to appearance of abnormal cell structure and functions as seen in triggering of the cancer cell formations.

See Fig.8   See Fig. 22

Fig-22

How Does Nature Work & Triggering of Normal & Abnormal Life Effects

To understand the triggering event of the normal and abnormal life effects, one must know about properties of B.B.Bs. At the time of the origin of the universe, all effects got created. The cause of all effects of the universe is **THOUGHT** expression. It is the first step and it is followed by **PROGRAMMING** or formation of programmed messages by code PCPs. This programmed message moves from higher centers to target B.B.Bs. it is called **INTERACTION**. Having received the messages, the mind and mass of the target B.B.Bs. work in a synchronized way so as to produced the effects as thought by a the higher center. If the thought expression by higher center is normal, the shapes, properties and laws produced by target B.B.Bs. would be normal and if the thought expressions are abnormal, the shapes, properties and laws would be abnormal. This is the basic concept of transmutation phenomenon. Finally what we observe is called **EFFECT**.

Appearance of new shapes. properties and laws is called **TRANSMUTATION**. The first three steps are collectively called **CCP**. During transmutation process if **CCP** is written, it does mean that unless the thought, programming and interaction take place, nature cannot transmutate. Transmutation phenomenon is seen in particles, atoms, molecules and even in cells. The basic steps of any transmutation remain the same except that the thought expressions differ.

See Figure. 22   See Figure.17

Triggering and Regulation of Cell Functions or Cell Physiology

Normal cell structure and function are triggered by atomic transcriptions or thought expressions by higher centers (B.B.Bs.) present inside the cell. These higher centers are found in **DNAs** and in the membrane. Higher centers in **DNA** express thoughts. The programmed messages (code PCP) are carried by photons from nucleus of the atom to electrons. Thus messages come on the surface of DNA molecule. These messages shift to mRNA during third transcription. **mRANs** carry the massages (code PCPs) from nucleus to cytoplasm where messages are translated by ribosome. Simultaneously, messages shift to peptide chains (enzymes &hormones) These enzymes and hormones carry the messages to target units (B.B.Bs.). Having received the messages, target units work accordingly and life effects like metabolism, cellular respiration, growth and structure formation, and other metabolic effects are observed. There is feedback of the effects to the higher centers. Thus triggering and regulation of metabolic functions are carried out.

Similarly, the higher centers of membrane express thoughts. These thoughts are carried by photons form nucleus to the electron of the atom. Thus messages (code PCPs) come on the surface of the membrane. Here they are modulated on action potential and thus action potential (second transcription) carries the messages from one part of the cell to the other part of the cell. This is called electrical effects of the cell. There is feed back of these effects to higher center and thus these effects are triggered and regulated.

Both the effects (metabolic and electrical) have there feed back not only to higher centers but also to highest centers via first transcription. So highest centers is well informed about all effects of the cell. Higher centers are not autonomous centers rather they are also under control of highest centers of the universe.

See Figure.23   See Figure.24

Mechanics of Life and Death (see Fig - 23 & 24)

Seed is alive (say having 1% of life activity) but hardly show any sign of life. It has low water content and exhibits virtually no metabolic activity. Such quiescent seeds can live for many years but germinate when soaked in water under suitable temperature and in presence of oxygen. Metabolic activity (anaerobic)are very low in seeds. Metabolic activities come to visually standstill as the seed coat becomes increasable impermeable to oxygen and moisture.

The first step in germination is IMBIBITION. Imbibition of water causes resumption of metabolic activities. Initially metabolism may be anaerobic (due to energy provided by the glycolysis) but it soon becomes aerobic as oxygen stats entering the seed.

What are life activities ?

DNA Activities

- Transcription - that leads to first anaerobic metabolism later aerobic metabolism. It is very very low in seed.

- Replication - it is nil in seed. Replication is the sign of life. If seed does not show replication phenomenon, it means for practical purpose it is dead.

During germination anaerobic metabolism is triggered and later it is shifted to aerobic metabolism. Replication is also triggered. The triggering of both the activities is onset of atomic transcriptions of replication as well as onset of transcriptions of aerobic metabolism. With the result messages come on the surfaces of DNAs and during 3rd transcriptions they are shifted to different mRNAs and finally they reach to different enzymes and hormones. These enzymes and hormones carry messages to target units. With the result we observe phenomenon of germination. The entire working has been depicted by line diagram. These atomic transcriptions are stimulated by water that why it is CONDITIONED STIMULATION of CCP. The percentage of thought expression increases with the time and we observe increase in the number of effects. At present we can say that plant is showing from 1 % to 20% or 40% of its effect till it reaches its maturity. At maturity all effects are exhibited by the plant and at that time we can say it is expressing 100% life atomic transcriptions. With the formation of new seed life atomic transcriptions once again reduced to 1% only. This cycle i.e going from 1% life effects or atomic transcription to 100% life effects and coming back to 1% again is being visible to us at present. When water is withdrawn, it leads to suppression of life thought expressions and death thought expressions are triggered with the result we observe death effects of plant.

CONCLUSION OF THE EXPERIMENT

The phenomenon of life effect is triggered by atomic transcription of life. Unless life atomic transcriptions are triggered, life effects are not visible. So life effects are nothing but higher thought expressions of basic building blocks. Phenomenon of death is triggered by death atomic transcription. At the time of death life thought expressions are inhibited and death thought expressions are triggered with the result we observe death effects. Water only stimulates life thought expressions that leads to triggering of different life activities (metabolic, replication and other electrical activities) inside the cell.

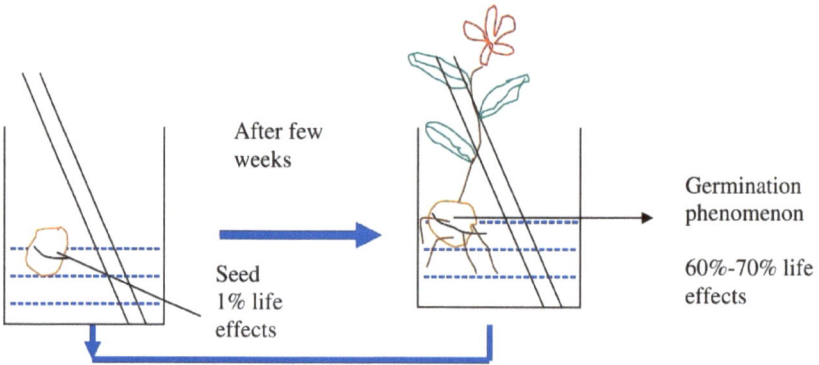

At the end of maturity the cycle is repeated.

Being a scientist, one must know how do life effects come about. Life effects are higher thought expressions of B.B.Bs. Formation of particles, atoms and molecules are due to lower thought expressions of B.B.Bs. But their higher thought expression lead to appearance of all life effects. One who knows properties of B.B.Bs. and atomic genetics, can understand how life effects are triggered. There is nothing like **SOUL**. It is a myth that when soul goes inside we get life and when it moves out we are dead. When thought expressions of life are suppressed and thought expressions, of death are triggered, we observe death effects. So life effects are basically triggered by atomic transcription occurring on B.B.Bs.

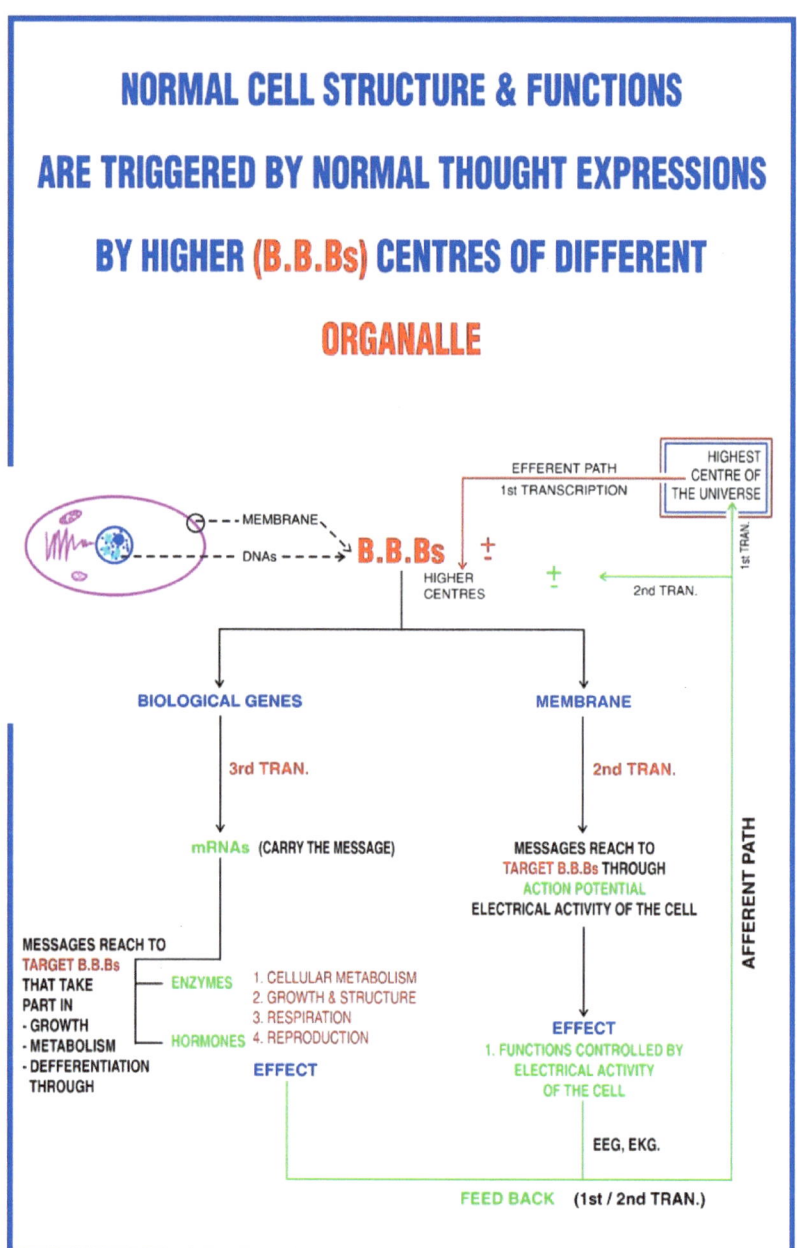

Fig-23

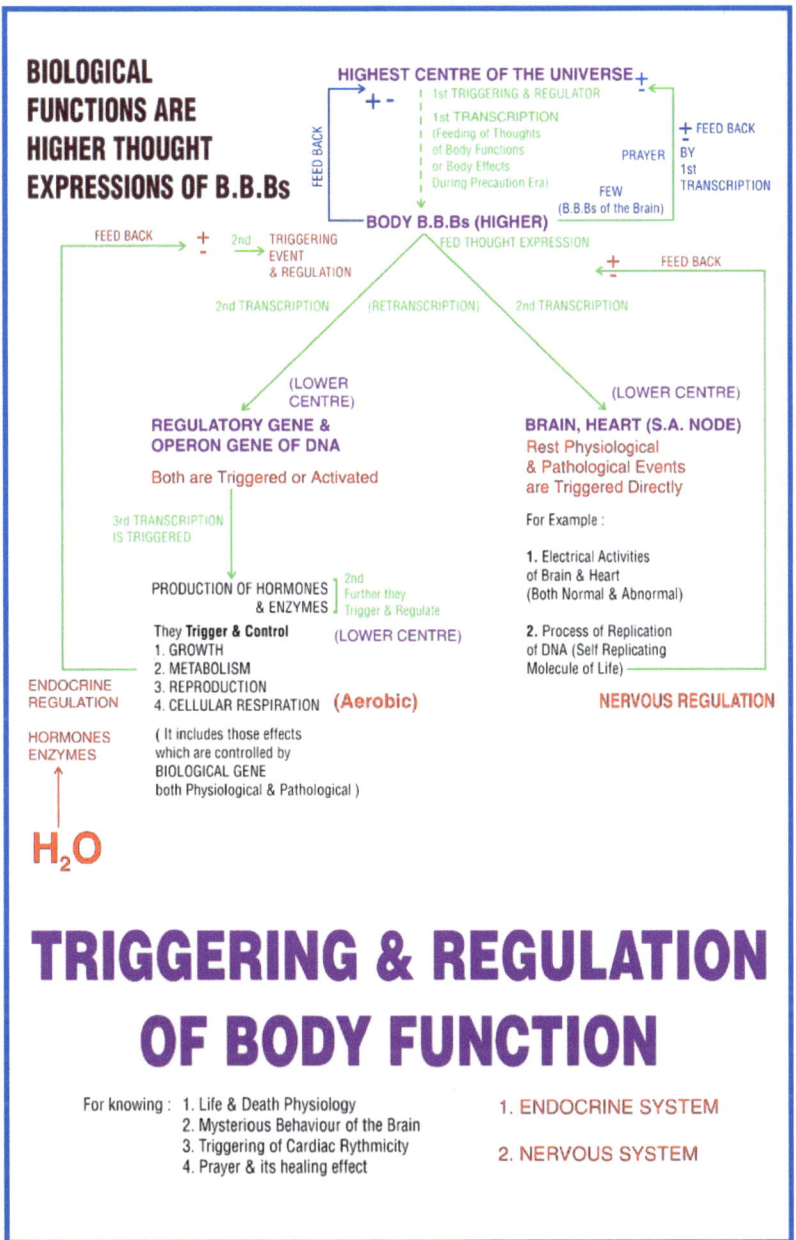

Fig-24

Cellular Oncogenes - I

Oncogenes have been found to induce cell transformation. Proto-Oncogenes are normal genes that regulate normal growth and differentiation. Proto-Oncogenes are the genetic segment carrying information as regard to the normal growth and differentiation.

Seeing the structure of proto-oncogene, it is made up of atoms and atoms are made up of B.B.Bs. There is a B.B.B. that works as higher center for regulation of normal growth and differentiation. Here thoughts of normal growth and differentiation are expressed. With the result there is formation of programmed messages of normal growth (code PCPs of normal growth) and differentiation (code PCPs of differentiation) These programmed messages are carried by photons from nucleus to electrons of the atom. Thus programmed messages are now on the surface of proto- oncogenes. Here during third transcription, messages are shifted to **mRNA**. **mRNA** carries the massages from nucleus to cytoplasm where during translation they shift to essential products. These essential products carry the messages to target B.B.Bs.or molecules which are responsible for normal growth and differentiation. Having received the messages, the mind and mass of the target B.B.Bs. work accordingly. As a result we observe normal growth and differentiation. Thus cell metabolism of normal growth and differentiation are triggered.

During cell transformation the thoughts that are expressing (ii & iii) for normal growth and differentiation are suppressed and thoughts for rapid growth and de-differentiation (v & vi) are triggered. As a result the programmed messages (code PCPs) of rapid growth and differentiation are there. They are carried by same route to essential products. These essential products carry the messages (v & vi) to target B.B.Bs. Having receive the messages, the mind and mass of target B.B.Bs. work accordingly. With the result instead of normal cell functions we observe rapid growth and de- differentiation or cancer.

See Figure. 25

Transformation of proto-oncogenes into oncogenes or Genetic Damage

Before the cell shows change of normal growth and differentiation to rapid growth and de- differentiation, it shows transformation of proto-oncogenes into oncogenes or genetic damage. The normal genetic configuration or proto-oncogene is triggered by normal genetic configuration thoughts (i). With the result there is formation of messages of normal genetic configuration (code PCPs normal genetic configuration). When these messages reach to target B.B.bs of DNA molecule, they (target B.B.Bs) start showing normal genetic configuration of DNA. Before the transformation of the cell, normal genetic configuration thoughts (i) are suppressed and abnormal genetic configuration thoughts (iv) are triggered. With the result abnormal genetic configuration messages (code PCPs abnormal genetic configurations) are formed. These messages are carried to proto-oncogenes. Having received the messages by the target B.B.bs of proto-oncogenes, the proto-oncogenes transform into oncogenes or we observe Genetic Damage. This is the first step **(mutation)** of phenomenon of carcinogenesis.

Cellular Oncogenes - II

Shifting of thought expressions (i, ii & iii) to (iv,v & vi) could be conditioned or unconditional. It does mean that thought expressions could be triggered either from outer stimulus (conditioned) or by self stimulus (unconditioned). Outer stimuli are physical and chemical carcinogens and virus etc. While self stimulus is specific genetic background. Outer stimuli carry the programmed thought of shifting (that is why they are called carcinogens). After coming in contact with normal cell, they send messages of shifting to higher center of proto-oncogene that regulate normal growth and differentiation. Say after years, higher center shifts the thought expressions from normal (i, ii & iii) to abnormal (iv, v & vi). The time period is in terms of years because higher centers were informed about there role in precreation era by highest center of the universe. Had highest center fed time of shifting in hours or days, it would have shifted in hours or days. It would be shown in the section of atomic genetics that how feeding of information was done in pre creation era.

See Figure.26   See Figure. 27

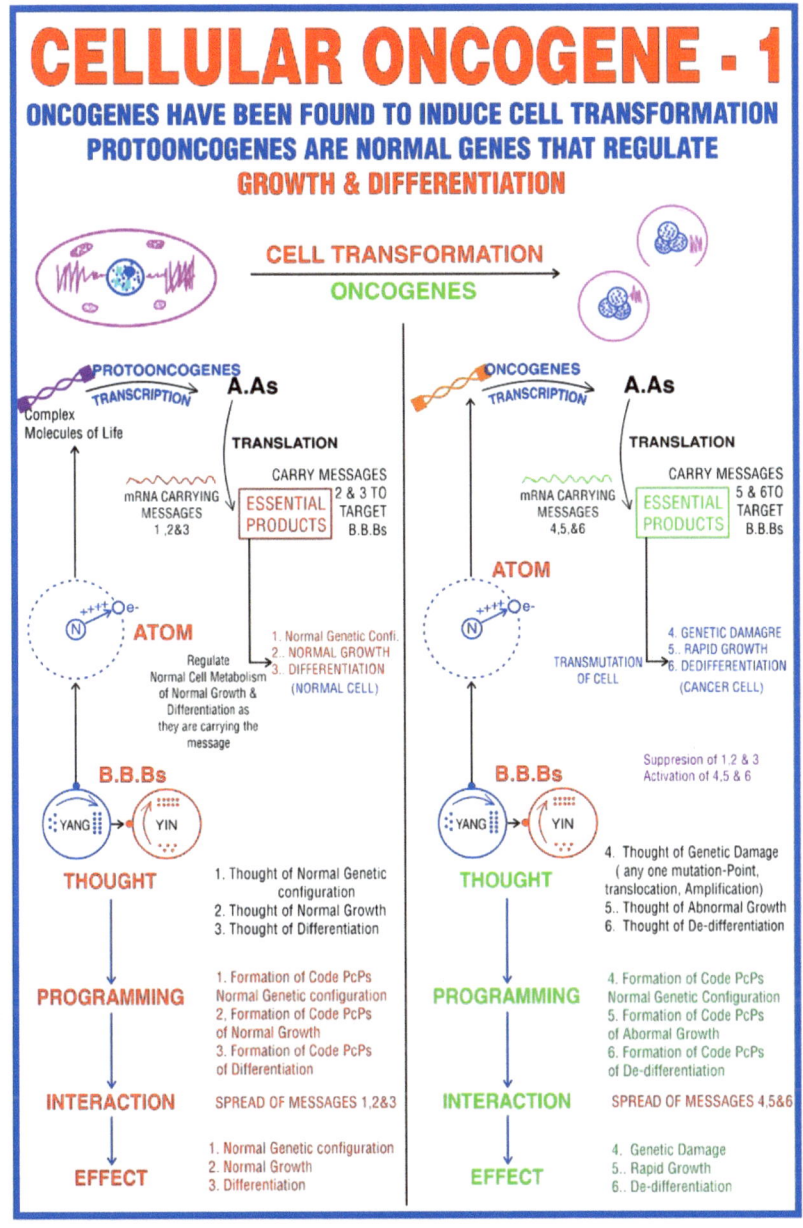

Fig-25

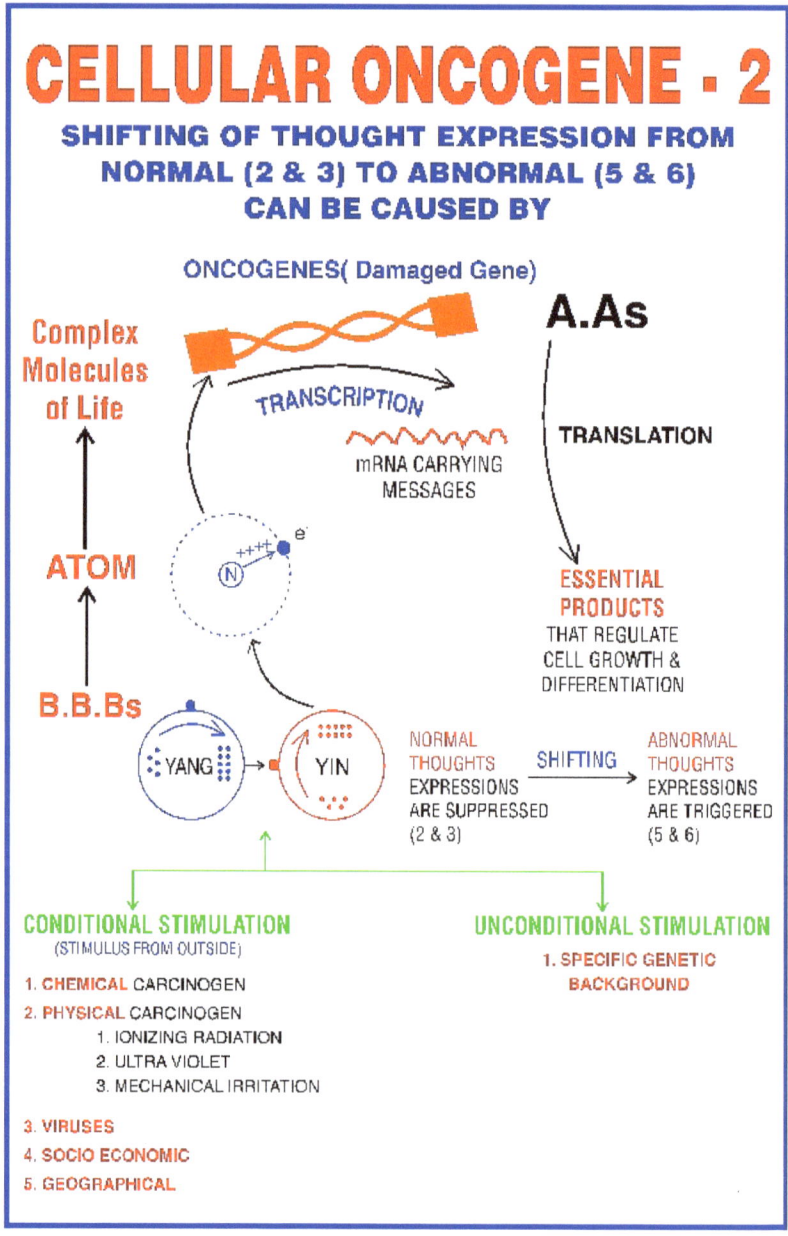

Fig-26

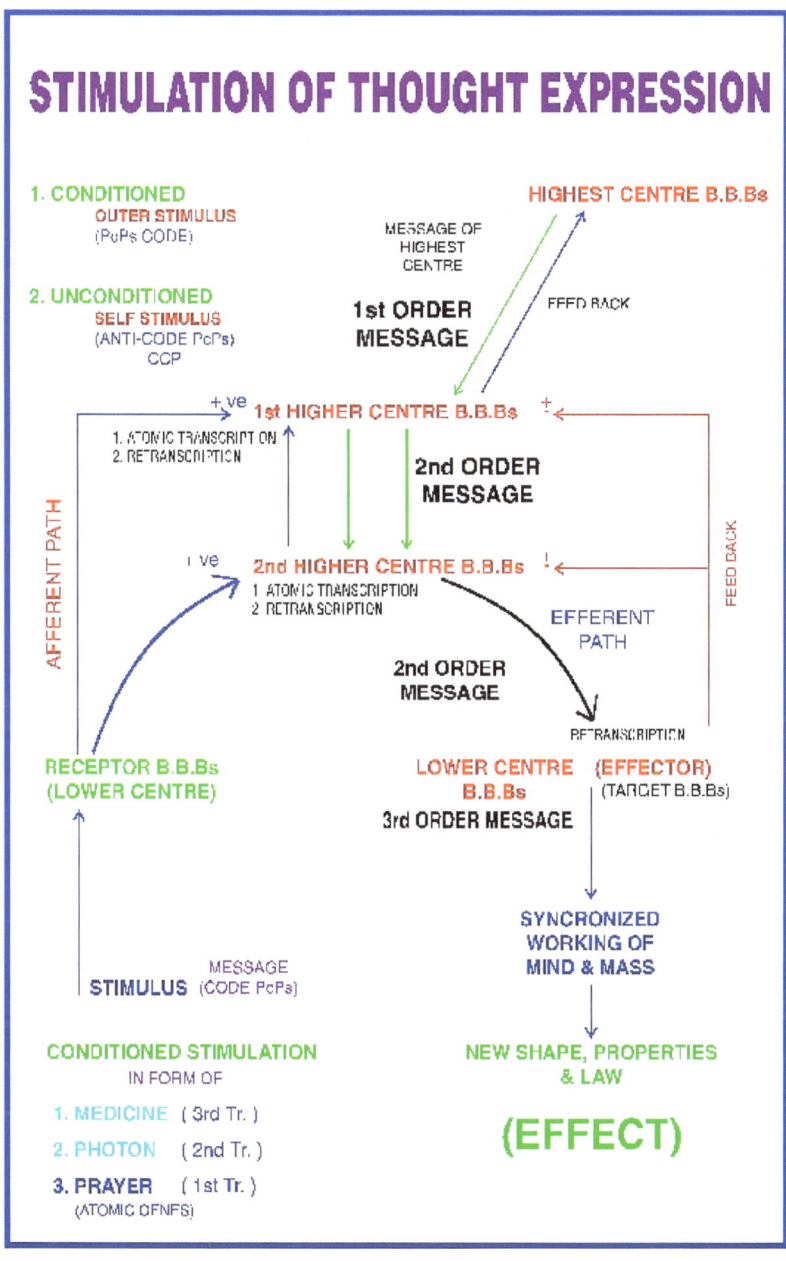

Fig-27

Atomic Genetics and Genetic Damage
(see Fig-22) [9]

Having read Atomic genetics and Basic etiology of the cancer, now we discuss how atomic genetics trigger molecular effects like genetic damage etc during carcinogenesis. ATOMIC GENETICS is a new branch of science in which we study about Laws, PROPERTIES and FUNCTIONS of atomic genes. Now I shall highlight the laws on which atomic genes work. As we have seen that Gregor Mendel had made three laws of inheritance known as

- The principle of Dominance— Mendel therefore concluded that what were transmitted from parent to offspring were discrete factors. Each factor contained information about the form of the trait. The factor associated with the form which was expressed in the hybrid offspring (F1) was **DOMINANT**. For example, the factor for yellow seed color was a dominant factor. The factor associated with the form which remained hidden in the F1 but reappeared in the F2 was **RECESSIVE,** Thus the factor for the green seed color was recessive. Mendel's factor is now recognized as the gene.

  In atomic genetics this law is interpreted like this —- During atomic transcription or thought expressions few thoughts are expressed and these are called dominating thoughts and rest thoughts which are not expressed are called recessive thoughts. Thus particles, atoms and molecules show only those properties which are triggered by dominating thought expressions. During transmutation, dominating thoughts get recessive while recessive thoughts get dominant. Single effect or property is triggered by single dominating thought expression or atomic transcription.

  For example– Normal genetic arrangements are triggered by normal arrangement genetic dominating thought expressions while genetic damage is triggered by abnormal arrangement genetic dominating thought expressions. Abnormal arrangement genetic thought expression is triggered by carcinogens only after normal arrangement genetic thought expressions get recessive. Or we can say normal genetic arrangement are triggered by normal arrangement genetic thought expressions and when carcinogens come in contact with the cell, they shift the thought expression from normal to abnormal genetic arrangement. With the result we observe genetic damage. Thus carcinogens suppress dominating normal arrangement genetic thoughts and they trigger abnormal arrangement recessive genetic thoughts. With the result genetic damage is seen.

- The Principle of Segregation—The principle of segregation states that allele pairs separate or segregate during gamete formation, and the paired condition is restored by random fusion of gametes during fertilization. In atomic genetics it will be interpreted like this — There is separate atomic transcription for separate effects. So if there are hundred effects, they all are triggered by hundred separate thought expressions or atomic transcriptions.

  For example— Genetic damage is triggered by separate thought expression and rapid growth is triggered by separate thought expression and dedifferentiation is triggered by separate thought expression. And all these thought expressions are triggered by carcinogens. But the timing of thought expressions is different. Genetic damage thought expression is first to trigger and later rapid growth and dedifferentiation thought expressions.

- The Principle of independent assortment—- The principle of independent assortment states that if we consider the inheritance of two or more genes at a time, there distribution in the gametes and in the progeny of subsequent generations is independent of each other. In atomic genetics it will be interpreted like this— In one phenomenon, if two or more than two transcriptions or thought expressions are expressed that will give rise to two or more than two effects, **it does not mean that there expression is DEPENDENT ON EACH OTHER.**

For example— In phenomenon of carcinogen sis, there is effect of genetic damage, there is effect of rapid growth and there is effect of dedifferentiation. All these effects are triggered by separate thought expressions. The simultaneous expression of these atomic transcriptions is independent of each other. Or these thoughts expressions are the part of the carciogenesis phenomenon but their expressions are independent of one another. **It does mean that for rapid growth thought expression, genetic damage thought expression is NOT essential.** Having informed about the laws and working of atomic genes, Now we discuss phenomenon of carcinogen sis, which is triggered by, outer stimuli like physical, chemical carcinogens or virus or they are self stimulated i.e. hereditary factors.

When Genetic damage is not essential for cancer transmutation then what is the role of Genetic damage? **Why do carcinogens trigger genetic damage first? How do carcinogens trigger carcinogenesis?**

- CARCINOGENS — Carcinogens carry a programmed thought of shifting. Having come in contact with the normal cell, they send messages of shifting to all higher centers **(BASIC BUILDING BLOCKS)** that are involved in thought expressions to give all effects of normal structure and functions of the cell including normal chromosomal configuration. Unless normal cell growth regulation gets damaged, carcinogens cannot trigger carcinogen sis.

- PROTO- ONCOGENES OR ONCOGENES —- Proto-oncogenes are the genes that express normal thought expressions; with the result we observe normal **GROWTH AND DEFERENTIATIONAL.** While oncogenes are same proto-oncogenes but they express abnormal thought expressions with the result we observe abnormal effects like rapid growth and de-differentiation. It is the carcinogens that transmutate proto-oncogenes into oncogenes. **(ONLY FUNCTIONAL CHANGE NO STRUCTURAL CHANGE EXCEPT DAMAGE).** Oncogenes (genes that are involved in cancer transformation) are **highly homologous** to cellular genes involved in normal growth and control (Proto oncogenes). It is the carcinogens that trigger phenomenon of carcinogenesis in oncogenes.

- GENETIC DAMAGE—- Point mutation, chromosomal translocation, and gene amplification, these are the three types of genetic damage effects seen in ras proto-oncogenes, c-fms gene (Point mutation) myc proto-oncogenes, bcl-2 proto-on cogenes, ab1 proto oncogenes (Translocation), N-myc and L-myc proto oncogenes (Gene amplification). The genetic damage is triggered by carcinogens. It is the first step of carcinogen sis.

It just simply shows that normal genetic chromosomal structure or configuration has been transmutate to abnormal chromosomal structure or configuration by carcinogens by shifting the normal thought expressions to abnormal thought expressions. Normal thought expressions were responsible for normal configuration of chromosomes and abnormal thought expressions are responsible for abnormal chromosomal configuration.

It also shows that scientists must think that some thing other than molecular changes that is occurring inside oncogenes that is the basic etiology of the cancer. This effect has concentrated the scientists to think inside proto-oncogenes, which will be responsible for rapid growth and dedifferentiation. Genetic damage shows that basic etiology is not genetic damage rather it is inside proto oncogenes which have been transformed into oncogenes (Damaged gene onc myc, fos, jun)

GENETIC DAMAGE HAD TRIGGERED THE DAMAGE OF NORMAL CONTROL OF CELL DIVISION AND THUS NORMAL CELL PHYSIOLOGY OF REGULATION GETS HAMPERED. This is observed by one of the alteration or damage in normal cell growth regulation process. (Fig-c)

- ONCOGENES— Oncogenes myc, fos, jun are responsible for rapid growth and dedifferentiation. Their proto-oncogenes were responsible for normal growth and differentiation. So what had happened inside oncogenes that triggered carcinogen sis. (Fig-25) It is the carcinogens that had shifted the normal thought expression to abnormal thought expression inside myc, fos, jun not the genetic damage.

Genetic damage occurred first and activation of oncogenes was followed by it. Both the effects are triggered by carcinogens, but the time period was different.

Apart from stopping normal cell growth regulation by alteration of transcription factors, they also trigger thought expressions of rapid growth and de-differentiation.

Role of myc, fos, jun oncogenes- Mitotic cycle is regulated by a family of genes whose products are localised to nucleus, where they control transcription of growth related genes. Not surprisingly, therefore, mutation-affecting genes that encode nuclear transcription factors are associated with malignant transformation. It means they not only affect nuclear transcription factors but also they trigger phenomenon of carcinogen sis. A whole host of oncoprotein, including products of the myc, myb, jun, and fos oncogenes have been localized to the nucleus. Of these the myc gene is the most commonly involved in human tumors.

Role of myc oncogene (Fig-a)

Apart from stopping normal cell growth regulation by alteration of transcription factors, they also trigger thought expressions of rapid growth and de-differentiation. Both the processes are triggered by carcinogens.

MUTATION (genetic damage) should not be correlated with **MALIGNANT TANSFORMATION** (rapid growth and de-differentiation) as mutation causes damage of normal cell growth regulation by damaging transcription activation. This message is fed back (red arrows) to higher centers. Both are triggered by separate thoughts and their simultaneous expression is independent of each other i.e. for malignant transformation mutation is not necessary

(Fig-a)

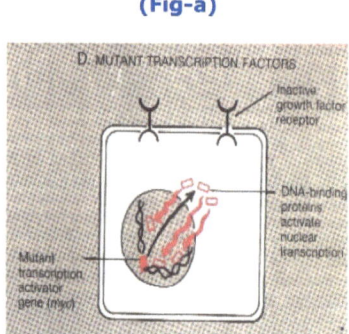

- PROTO ONCOGENE AND ONCOGENE PRODUCTS (ESSENTIAL PRODUCTS –Proto oncogenes products (from mRNA to essential products) were carrying normal messages of growth and differentiation with the result we observe normal growth and differentiation. While oncogene products (from mRNA to essential product) were carrying abnormal messages or rapid growth and dedifferentiation that is why we observe phenomenon of carcinogen sis i.e. rapid growth and de-differention. CARCINOGENS (not by GENETIC DAMAGE) cause this shifting of thought expressions from normal to abnormal.

Normal molecular events in cell growth regulation which are damaged by damaged genes are following—- (Fig-b) [9]

LIGAND- (GROWTHFACTORS) RECEPTOR BINDING

GROWTH FACTOR RECEPTOR ACTIVATION

SIGNAL TRANSDUCTION AND SECOND MESSENGER

TRANSCRIPTION FACTORS

PLACES OF ACTION OF DIFFERENT PROTO-ONCOGENES IN REGULATING MITOTIC CYCLE

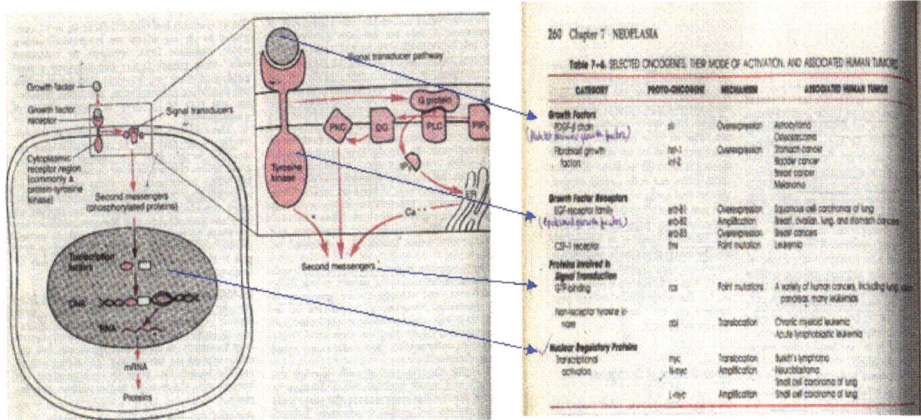

Fig-b

It must be noted that increased growth factor production is not sufficient for neoplastic transformation. Same is true for Increased growth factor receptors formation, alteration in signal transduction and second messengers and alteration in nuclear regulatory proteins. Either of these alteration stops normal regulation of cell division. This altered messages (feed back) are sent to higher center (B.B.Bs), which were involved in damage of normal cell growth regulation. Having stopped the normal cell division regulation, carcinogens trigger activation of Oncogenes. These activated oncogenes (myc, fos, jun) trigger phenomenon of carcinogenesis by expressing abnormal thought expressions of rapid growth and de-differentiations. With the result phenomenon of carcinogenesis is observed.

The regulatory mechanism of normal growth and differentiation has been damaged by damaged gene. This message has been fed back (red arrows) to higher centers(B.B.Bs) that were involved in damage of normal growth and differentiation. (Fig-c)

GROWTH FACTORS

Normal amount of growth factors (PDGF -BETA CHAIN, Fibroblast growth factors) shows normal regulation of cell division process while increased amount

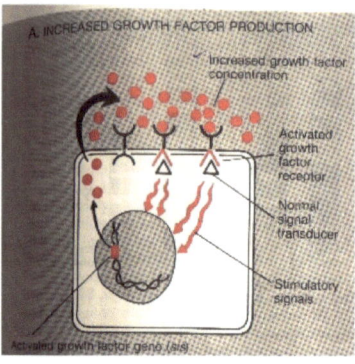
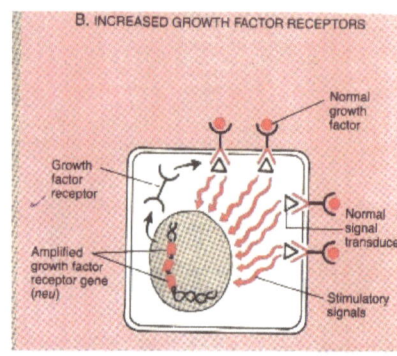
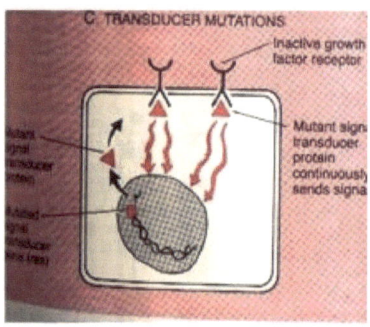
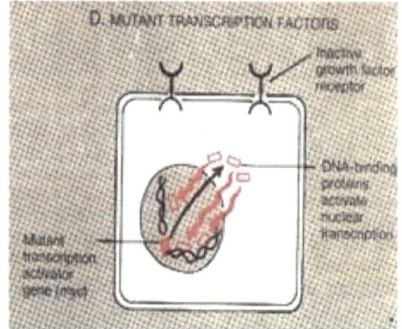

Fig-c

of growth factors means damage in normal cell growth regulation. Proto-oncogenes sis, hst-1, Int-2 (over expression) are responsible for this transmutation in cancers.

GROWTH FACTOR RECEPTOR ACTIVATION

Normally there are few growth factor receptors (EGF receptor family, CSF-1 receptor) that take part in normal cell division (mitosis). In carcionogenesis their number get increased. Proto-oncogenes

erb-B1, erb-B3 (over expression), erb-B2 (amplification) and fms (point mutation) are responsible for this transmutation in cancers. It shows normal cell growth regulation is damaged.

SIGNAL TRANSDUCTION AND SECOND MESSENGER

Normally the surface messages are shifted to nucleus for cell division and molecules involved are GTP - binding, No receptor tyrosin kinase. Alternation in the signaling pathway means alteration in the messages carried by them or damage of normal cell regulation process. Proto-oncogenes ras (point mutation), abl (translocation) are responsible for this transmutation in cancers.

TRANSCRIPTION FACTORS

Normally transcriptional activators (MAP kinase, CA++, calmoduline) control the transcription and growth related genes and thus activates transcription and mRNAs are formed. These mRNAs also carry messages of normal growth and differentiation. With the result we observe mitotic cycle of normal growth and differentiation. Alteration in transcription factors leads to damage of normal cell growth regulation. Proto oncogenes myc (translocation). N-myc (amplification) and, L-myc (amplification), fos, jun are responsible for this transmutation in cancers.

Large number of cellular genes are divided into

1. EARLY GROWTH-REGULATED GENES (c-fos, c-jun, c- myc), whose mRNA increase well before mid-G1 of the cell cycle and which are induced in the absence of protein synthesis.

2. LATE GROWTH REGULATED GENES, whose mRNA start to increase in mid G1, or even at the G1-Gs boundary and which are dependent on protein synthesis (Fig-d).

Among the growth regulated genes are a number of proto-oncogenes in which mutations may be associated with malignant transformation. Some, such as myc, fos, and jun, code for transcription factors and are involved in the regulation of DNA synthesis and possibly, cell division.

IS MUTATION IN mys, fos, jun ASSOCIATED WITH MALIGNANT TRANSFORMATION ?

MUTATION IN myc, fos, jun IS NOT ASSOCIATED WITH MALIGNANT TRANSFORMATION — GENETIC DAMAGE (MUTATION) IS A SEPARATE EFFECT WHILE TRIGGERING OF CARCINOGEN SIS IS A SEPARATE AND BOTH ARE TRIGGERED BY CARCINOGENS BUT

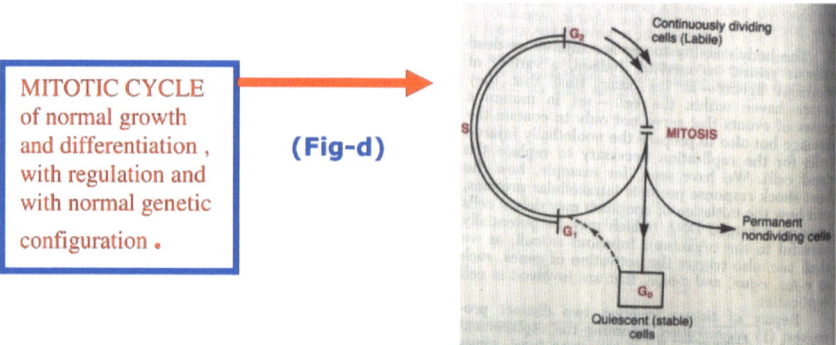

MITOTIC CYCLE of normal growth and differentiation , with regulation and with normal genetic configuration .

(Fig-d)

THEIR TIMING PERIOD IS DIFFERENT. HAD THERE BEEN NO GENETIC DAMAGE (MUTATION) IT WOULD HAVE BEEN DIFFICULT TO LOCATE THE SITE OF TRIGGERING OF PHENOMENON OF CARCINOGENESIS INSIDE myc, fos jun. Mutation and malignant transformation are triggered by separate thoughts. Their simultaneous expression give rise to mitotic cycle of rapid growth and de-differentiation without regulation with genetic damage. Their simultaneous expression is independent of each other it does mean that for malignant transformation, mutation is not essential.

Damage shows that some thing other than damage (expressing thoughts of carcinogenesis) is also occurring inside these genes which were responsible for carcinogen sis.

Carcinogens not only trigger transmutation of proto oncogene into oncogene but also they trigger phenomenon of carcinogen sis by shifting of thought expressions from normal to abnormal.

- PROTEIN PRODUCTS OF ONCOGENES. Oncogenes encode proteins called oncoproteins, which resemble the normal products of proto-oncogenes, with the exception that oncoproteins are devoid of important regulatory elements and their production in the transformed cell does not depend on growth factors or other external signals.

- CANCER SUPPRESSOR GENES—- Rb, p53, APC, WT-1, DCC, NF-1, NF-2,VHL are responsible for suppressing the rapid growth in normal cell. Carcinogens again suppress their thought expressions and thus carcinogen sis is enhanced. Again it is not the genetic damage that inactivates tumor suppressors genes. **IT IS THE CARCINOGENS THAT SHIFTS ALL NORMAL THOUGHT EXPRESSIONS TO ABNORMAL THOUGHT EXPRESSIONS NOT THE GENETIC DAMAGE.** [9]

- STIMULATION OF THOUGHT EXPRESSION — Shifting of thought expressions from normal to abnormal could be triggered by either from outer stimuli i.e. CONDITIONED STIMULATION of thought expression, which is caused by physical and chemical carcinogens or viruses etc or it is self stimulate i.e. UNCONDITIONED STIMULATION of thought expression, which is caused by hereditary factors. (Fig-26 & 27)

CONCLUSION

Cancer is supposed to a multi step phenomenon and the basic etiology does not lie in genetic damage or molecular basis. Cancer is multi step phenomenon but carcinogens trigger transmutation occurring in each step by sending the messages of shifting to each higher center (**BASIC BUILDING BLOCKS**) involved in multi step. The basic etiology is not molecular genetics damage rather shifting of thought expressions from normal to abnormal thought expressions at the level of **BASIC BUILDING BLOCKS** that constitute that **molecule**. With the result the molecules show changes in structures, functions and laws.

Phenomenon of transmutation means change in shape, properties and laws. All three are triggered by separate thought expressions. In carcinogen sis all three thoughts are transmutating. It could be chromosomal structural changes, It could be cellular structural changes or it could be biochemical changes, It could be staining changes, It could be any other, which is not present in normal cell structure and functions. So the basic cause in all these changes is shifting of normal atomic transcription to abnormal atomic transcription and carcinogens trigger that. Carcinogen sis is supposed to be a multi factorial disease. Yes, it is multi factorial. But what is **COMMON** in all factors. All factors carry programmed thought of shifting from normal thought expressions to abnormal thought expressions. What is common in all cancer transmutation? It is the shifting of thought expressions from normal to abnormal thought expressions and carcinogens trigger that or it is self stimulated (hereditary factors).

LAWS OF INDEPENDENT ASSORTMENT STATES THAT IT IS AN ILLUSION THAT GENETIC DAMAGE TRIGGER ACTIVATION OF

ONCOGENES AND FURTHER THEY TRIGGER ONCOGENESIS.

It is the basic building blocks (higher centers), which are the basis of molecular alterations, and the fundamental characteristics are shifting of thought expressions from normal to abnormal that is shared by all malignant tumors. The basic principles that govern carcinogen sis, are the laws made by **GREGOR MENDEL** i.e. the laws of INHERITANCE. Genetic damage is an effect not the cause of cancer. THE CAUSE OF CANCER IS SHIFTING OF THOUGHT EXPRESSIONS FROM NORMAL TO ABNORMAL, WHICH IS CAUSED BY EITHER OUTER STIMULI, OR THEY ARE SELF-STIMULATED.

Without shifting the thought expressions from normal to abnormal, neither one can have genetic damage, nor rapid growth or de- differentiation. (Fig 28)

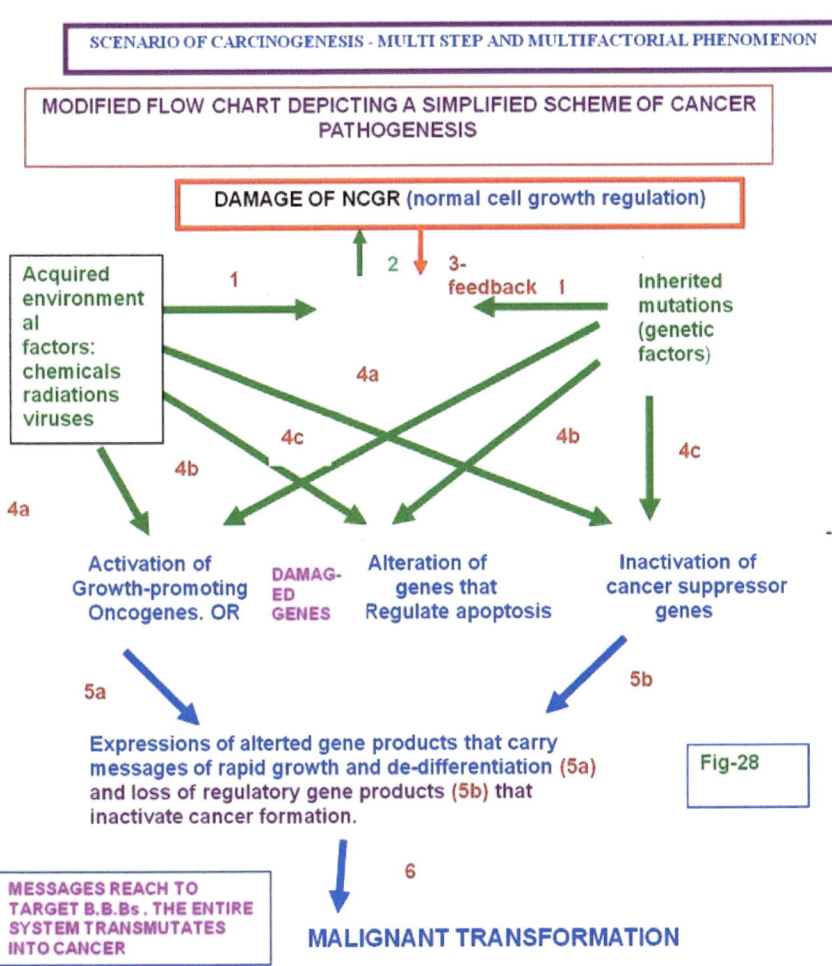

Fig-28

Atomic Genetic Engineering as Adjuvant Therapy & Message Formation in the Brain

In atomic genetic engineering we use our basic power i.e. power of B.B.Bs. Our B.B.B. (higher center) talks with highest center of the universe by sending the message by first transcription. Till today nobody knows how does the brain generate thoughts. I am going to tell you that mystery too. In the frontal lobe the neurons are responsible for thought generation. In the neuron there is electrical activity called pacemaker activity which is occurring between dendrites and the body of the neuron. The membrane of the cell is made up of atoms and atom is made up of B.B.Bs. At the level of B.B.B. say thought of **'O GOD HELP ME'** is expressed. As a result programmed messages of O GOD HELP ME (code PCPs) are formed. Out of three programmed messages, one is carried by atomic genes to highest center of the universe. It is called **THOUGHT RAY**, which is made up of pure atomic genes, and then the message goes through phenomenon called first transcription. They come out from brain directly. The other two messages are carried by photons from nucleus of atom to electrons. Here they are modulated on electrical activity of the cell called pacemaker activity. Further they are modulated on actions potentials going towards **REALIZING CENTER** situated in brain stem (RAS) and towards speech area situated in the Frontal area via RAS. Target B.B.Bs. of the realizing center finally realizes thought effect of O GOD HELP ME. While from speech area message goes to motor cortex via RAS and from there to vocal cords and finally it comes out as a speech effect of O GOD HELP ME. In layman's terminology formation of the thought ray means PRAYER. The details would be given in next section of brain and atomic genetics along with the other mysteries of the brain. see Figure.30.

The message goes to highest center of the universe where it is realized and it is accepted, the highest center sends two messages to B.B.Bs. working as higher center in cancer cell. These messages are message of inhibition of abnormal thought expression and message of activation of normal thought expression. Having received the messages, higher center stops expressing the abnormal thoughts and it starts expressing the normal thoughts. As a result, there are no more abnormal programmed messages and in place of that normal programmed messages are there. Now the messages have shifted from abnormal (iv,v & vi) to normal (i, ii & iii). This shifting of thought expression is called **ATOMIC GENETIC ENGINEERING.**

The changed messages reach to target B.B.Bs. through same route. Having received the changed messages, target B.B.Bs. stop expressing the previous programming and they start expressing the normal programming. As a result the cancer cells transmutate into normal cells. It is the prediction of theory. See Figure.29

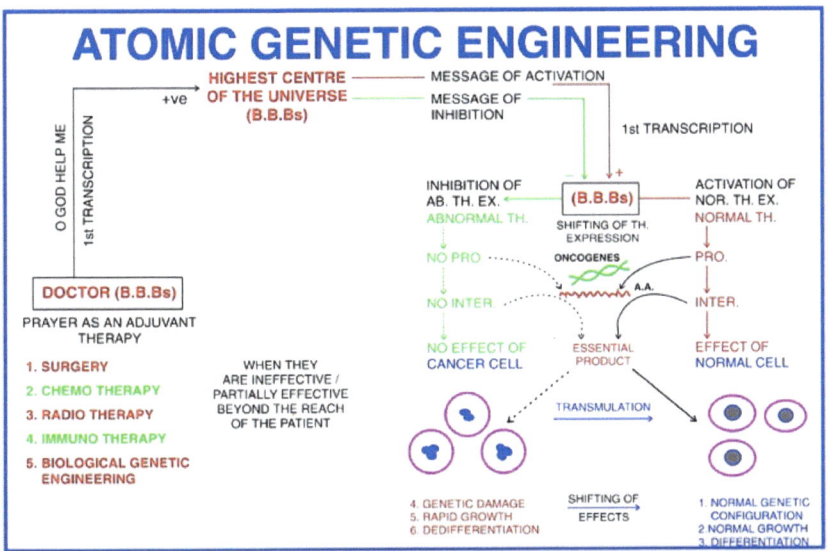

Fig-29

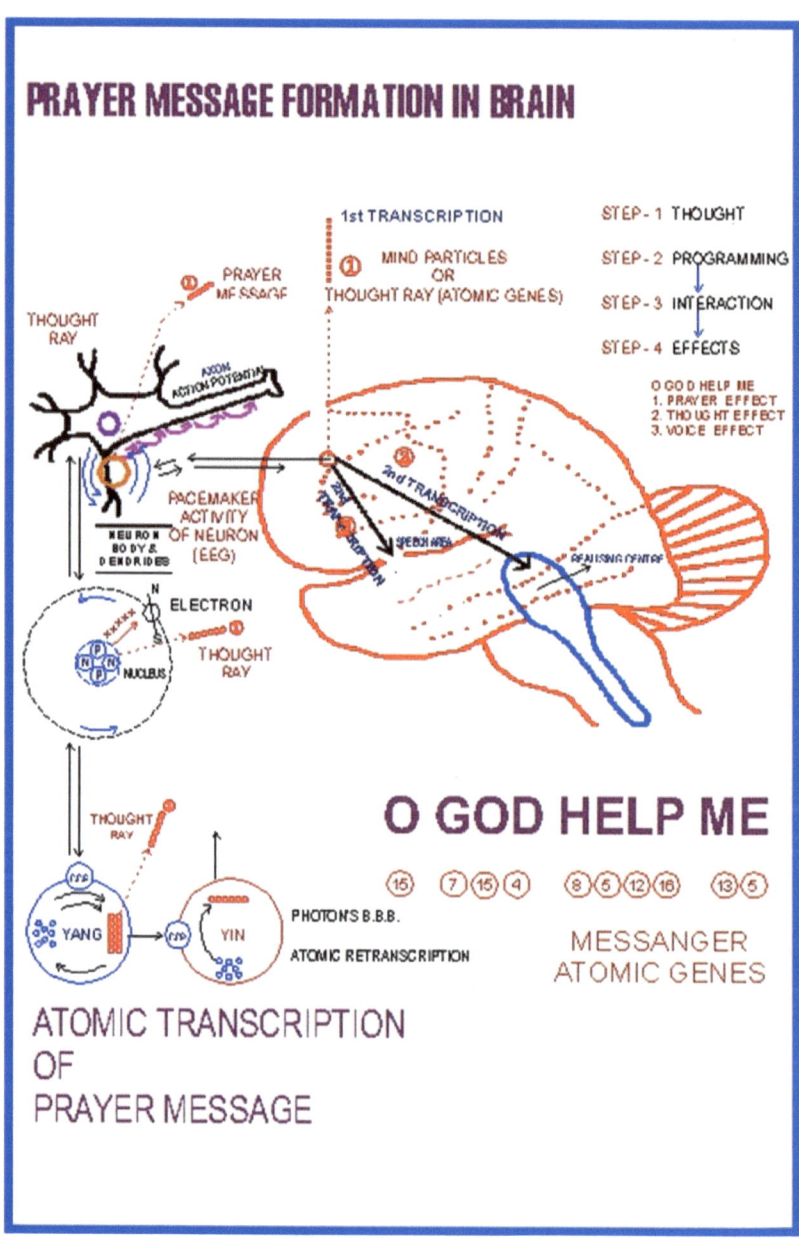

Fig-30

Prayer and Growth Dynamics of Cancer

Participatory science advocates PRAYER as an adjuvant therapy to improve cure rate (5years, 10years and so on) in all stages (stage one to stage four) of the cancer. Seeing the growth rate of neoplasm, it takes 30 exponential divisions to produce 1cm nodule (1 billion cells). At 45 exponential divisions the patient is apt to be dead from the sheer bulk of the malignant tumor. Based on growth dynamics most tumors have been present in the body for at least 1 year and may for long as 10-15 years prior to the clinical detection. Thus it appears that there is a long period of time between the inception neoplastic transformation and the development of clinical cancer. Unfortunately, one of the great difficulties in the present staging method is that inability to detect sub clinical microscopic metastatic lesions. Many patient who are treated for apparently localized cancers already have disseminated metastasis. (For example, about one half of those patients who have cancer of the breast and who undergo mastectomy have sub clinical distant metastasis at the time of operation.) The localized lesions (T1N0M0) has crossed 30 exponential divisions to produce 1cm nodule. What had happened to cancer cells between 1 to 29 exponential divisions ? The answer is some time few cells migrate from the primary site to reach secondary and where they proliferate faster to give secondary lesion first (T0N1M0). Thus the initial presentation of the tumor may be at the distant from its origin. In fact, the primary neoplasm giving rise to the metastasis may have regressed completely and may never be detected in some neoplasm. The most common metastatic sites for unknown primary are cervical and supraclavicular lymph nodes, lungs,liver, bone and brain. Local recurrence of the cancer following surgery may be due to incomplete removal or spillage of the cells into the operation area. Believing all clinical stages are highrisk group, participatory science advocates prayer as an adjuvant therapy to conquer cancer in all stages of the cancer. The theory predicts that if prayer is tried honestly (intercessory prayer - Archives internal medicine- JAMA, vol 159, No- 19, 25th Oct. 1999- STUDY - PRAYER HELPS CARDIAC PATIENTS by William S. Harris, PhD), the cure rate will be improved (in %) in all the stages of the cancer irrespective of growth dynamics of cancer. In prayer (**ATOMIC GENETIC ENGINEERING**) our B.B.B(Basic Building Blocks) talks with highest center of the universe to suppress abnormal thought statements and to trigger normal thought statements and thus the left out cancer cells could be transformed to normal cells making the recurrence zero.

(Fig-29)

Predictions & New Observations of the New Theory

The theory further predicts that if man wants to tame cancer, man has to learn atomic genetic engineering as adjuvant therapy. In atomic genetics engineering, our B.B.B. talks with highest center of the universe via first transcription to shift abnormal thought expressions to normal thought expressions. Thus the left out cancer cells could be transformed into normal cells making the recurrence rate zero. At present 11% success rate has been achieved using atomic genetic engineering (PRAYER) as adjuvant therapy in cardiac cases by Mid American Heart Institute, U.S.A. and it is published in JAMA (Fig-31) [6].

FINAL STAMP OF SUCCESS TO THE NEW THEORY

A Randomized, Controlled Trial of the Effects of Remote, Intercessory Prayer on Outcomes in Patients Admitted to the Coronary Care Unit

aainfoaainfo*William S. Harris, PhD; Manohar Gowda, MD; Jerry W. Kolb, MDiv; Christopher P. Strychacz, PhD; James L. Vacek, MD; Philip G. Jones, MS; Alan Forker, MD; James H. O'Keefe, MD; Ben D. McCallister, MD*

Context Intercessory prayer (praying for others) has been a common response to sickness for millennia, but it has received little scientific attention. The positive findings of a previous controlled trial of intercessory prayer have yet to be replicated.

Objective To determine whether remote, intercessory prayer for hospitalized, cardiac patients will reduce overall adverse events and length of stay.

Design Randomized, controlled, double-blind, prospective, parallel-group trial.

Setting Private, university-associated hospital.

Patients Nine hundred ninety consecutive patients who were newly admitted to the coronary care unit (CCU).

Intervention At the time of admission, patients were randomized to receive remote, intercessory prayer (prayer group) or not (usual care group). The first names of patients in the prayer group were given to a team of outside intercessors who prayed for them daily for 4 weeks. Patients were unaware that they were being prayed for, and the intercessors did not know and never met the patients.

Main Outcome Measures The medical course from CCU admission to hospital discharge was summarized in a CCU course score derived from blinded, retrospective chart review.

Results Compared with the usual care group (n=524), the prayer group (n=466) had lower meanSEM weighted (6.350.26 vs 7.130.27; $P=.04$) and unweighted (2.70.1 vs 3.00.1; $P=.04$) CCU course scores. Lengths of CCU and hospital stays were not different.

Conclusions Remote, intercessory prayer was associated with lower CCU course scores. This result suggests that prayer may be an effective adjunct to standard medical care .

JAMA--Arch Intern Med. 1999;Vol.159, No. 19 , 25 Oct. 1999 :2273-2278

(figure 31)

Comparative Study of Normal and Cancerous
Mitotic Cycle (Fig-32)

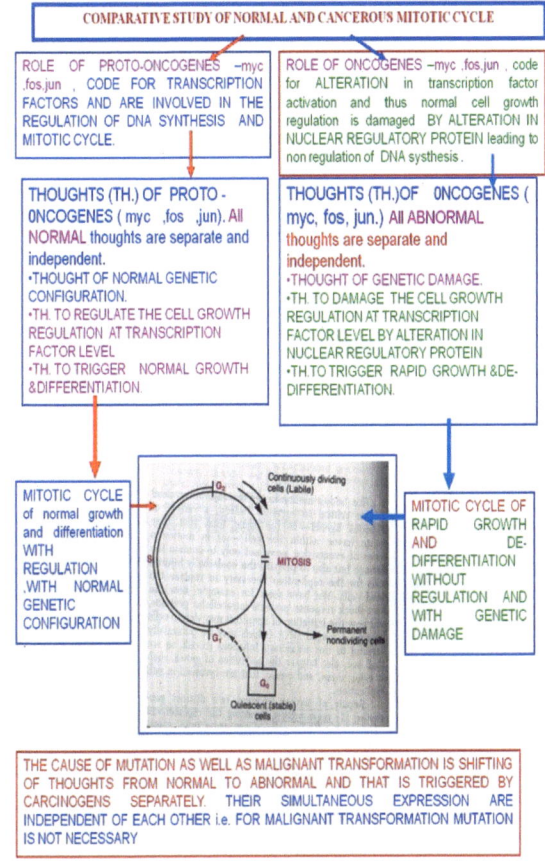

COMPARATIVE STUDY OF NORMAL AND CANCEROUS MITOTIC CYCLE

ROLE OF PROTO-ONCOGENES –myc .fos.jun , CODE FOR TRANSCRIPTION FACTORS AND ARE INVOLVED IN THE REGULATION OF DNA SYNTHESIS AND MITOTIC CYCLE.

ROLE OF ONCOGENES –myc .fos.jun , code for ALTERATION in transcription factor activation and thus normal cell growth regulation is damaged BY ALTERATION IN NUCLEAR REGULATORY PROTEIN leading to non regulation of DNA systhesis .

THOUGHTS (TH.) OF PROTO - ONCOGENES (myc ,fos ,jun). All NORMAL thoughts are separate and independent.
•THOUGHT OF NORMAL GENETIC CONFIGURATION.
•TH. TO REGULATE THE CELL GROWTH REGULATION AT TRANSCRIPTION FACTOR LEVEL
•TH. TO TRIGGER NORMAL GROWTH &DIFFERENTIATION.

THOUGHTS (TH.)OF ONCOGENES (myc, fos, jun.) All ABNORMAL thoughts are separate and independent.
•THOUGHT OF GENETIC DAMAGE.
•TH. TO DAMAGE THE CELL GROWTH REGULATION AT TRANSCRIPTION FACTOR LEVEL BY ALTERATION IN NUCLEAR REGULATORY PROTEIN
•TH.TO TRIGGER RAPID GROWTH &DE-DIFFERENTIATION.

MITOTIC CYCLE of normal growth and differentiation WITH REGULATION .WITH NORMAL GENETIC CONFIGURATION

MITOTIC CYCLE OF RAPID GROWTH AND DE-DIFFERENTIATION WITHOUT REGULATION AND WITH GENETIC DAMAGE

THE CAUSE OF MUTATION AS WELL AS MALIGNANT TRANSFORMATION IS SHIFTING OF THOUGHTS FROM NORMAL TO ABNORMAL AND THAT IS TRIGGERED BY CARCINOGENS SEPARATELY. THEIR SIMULTANEOUS EXPRESSION ARE INDEPENDENT OF EACH OTHER i.e. FOR MALIGNANT TRANSFORMATION MUTATION IS NOT NECESSARY

Fig-32

Highest Center of the Universe

This photo of "I" first God of symmetry phase is a representation of the highest center of the universe. The entire creation and destruction is under control of it. It is male part of androgynous form of "I", the Universal God. The entire creation and destruction are under control of highest center of the universe. The other names of highest center are Nat raj in Hindu, **TAO** in Taoism, ALLAH IN ISLAM and YAHOVA or PARMESHWAR or IN CHRISTIANITY. During atomic genetic engineering (prayer) our B.B.Bs. talk with highest center of the universe via first transcription. First transcription is the fundamental working and our basic power of the universe. The highest has power to change any earlier programming programmed by it during pre creation era.

See Figure.33

Fig-33

HIGHEST CENTRE OF THE UNIVERSE

This photo of "I" first God of symmetry breaking phase is a representation of the highest center of the universe. The entire creation and destruction is under control of it. It is male part of androgynous form of "I" first God of symmetry phase, the **ALMIGHTY** God. The entire creation and destruction are under control of highest center of the universe. Halos of fire represent the appearance of brightness (photons) in the universe by which universe can be visualized, while foot over the demon means destruction of the universe and both the process are under the control of highest center of the universe which is shown as **"I"** in the picture or male part of androgynous from i.e. **YANG** working as highest center of the universe.

Glossary

- PARTICIPATORY SCIENCE—— To study the nature by a new discipline in which realization and Common sense is the main tools other than observations. For example - as we realize that we work through phenomenon of thought similarly nature works through phenomenon of thought. As we realize that our walking is triggered by our mind similarly electron moves through triggering of its mind. As we interact through thought message system similarly forces interact through thought message system. Our behavior changes with change of thought similarly properties of the matter and energy change with change of their thoughts. We do research by principle of analogy or parallel. It is a new way to know unknown by the help of known knowledge. The known knowledge is either biological knowledge or mystical knowledge.

- ENERGIZED GRAVITONS—— Gravity force is mediated by the strings of particles (fermions) called Energized gravitons. During interaction (incoming energized gravitons) with unit system (apple) they give some energy to that system and thus the unit system gets accelerated. The out going energized gravitons are now at low energy level. Thus in every interaction with unit systems energized gravitons loose their energy and become low energy energized gravitons. Their energy (energy pool of the universe) content is never exhausted.

- SECONDARY BOSONS———- Rest three natural forces are mediated by strigs of particles (bosons) called secondary bosons.(gluons, vector bosons, Higgs bosons and photons).

- GRAVITON—— If.nature removes functional energy (F.E.) of energized graviton as well as the spin energy (B.E.-2 or Spin-2), the left particle is called graviton.

- PRIMARY BOSON—-If nature breaks secondary bosons, primary bosons will be created.

- MATTER B.B.Bs—- When nature breaks fermions into their B.B.Bs, the smallest mass unit which have got inertial property of absolute rest, is called matter B.B.B. (pure matter mass unit). The another name of this B.B.B is YANG. It is also called male part of androgynous form (unity of opposite).

- ENERGY B.B.Bs—— When nature breaks fermions and bosons into their B.B.Bs, the smartest mass unit that has got inertial property of motion is called energy B.B.B. The another name of this B.B.B. is YIN. It is also called female part of androgynous form (unity of opposite).

- ATOMIC GENES—- The conscious particles by virtue of which matter and energy attain transmutation phenomenon, properties, laws, interactions, message systems, behavior, life phenomenon, thought phenomenon, creation phenomenon is called atomic genes. They are the mind particles which conduct deterministic order of the universe.

- BASIC BUILDING BLOCKS—- The two types of smallest mass units of the universe having opposite properties (unity of opposite) which are divine in the sense that they talk with each other by phenomenon called atomic transcription and translation or thought statements. They have power to transmutate to form every bigger units of the universe.

- MIND AND MASS PART OF REALITY—— These B.B.Bs are made up of two realities. Firstly the mass part of reality that is objective reality by virtue of which bigger units attain shape or this gives shape to the bigger units (fermions and bosons) and that is why they occupy space. Secondly the mind (atomic genes) part of reality or subjective reality, by virtue of which bigger units attain different properties both physical and life sciences and laws both physical and life sciences laws.

- m of the statement $E = mC^2$—— It is also called impure matter particles (fermions) which have got spin properties.

- Pure m (matter) mass— Matter mass (smallest mass unit) which have got inertial property of absolute rest. It is also called matter B.B.B. or YANG.

- MASS—— The part of basic building blocks by virtue of which basic building blocks attain shape is called mass.

- INERTIAL MASS— The inertial mass of the bigger units (fermions and bosons) is by virtue of which they attain shape. In the other words number of B.B.Bs present in that unit system (fermions or bosons).

- GRAVITATIONAL MASS—- The gravitational mass is the mass of the bigger unit that is due to interaction between strings of energized gravitons and that unit.

- CCP (cosmic conscious particles) ATOMIC GENES—- It is thought script of the nature. It is similar to the DNA script of the biological world. It is made up of anti mind particles.

- Code PCPs(programmed conscious particles) —— These are messenger atomic genes. They are similar to the mRNA (messenger ribonucleic acid) of biological world. They carry messages (mind particles) from one B.B.B to another B.B.B.

- CPs (Conscious particles)—— These are translations (realizing recognizing and reacting) of atomic genes. The are similar to ribosome of biological world.

- Tachyons— Particles move more than velocity of light. The calculated velocity by participatory science is 305224Km/sec.

- HIGHEST CENTER OF THE UNIVERSE—— YANG or matter B.B.B or MALE part of androgynous form working as highest center of the universe. The prayer reaches to this center. It has got power of creation of the universe as well as destruction of the universe. It is symbolized as avatar of "I" First God of symmetry breaking phase..

References

1. D.N.VASUDEVA. Text Book Of Light, 7th edition, 1969; p. 598

2. Jayant V. Narlikar, Physics News- Vol-30, No- 3 & 4, Sept. & Dec, 1999; pp. 5-14

3. Fritjof Capra, The Tao Of Physics, 1989; p.153.

4. Fritjof Capra, The Tao Of Physics, 1989; p. 353.

5. Fritjof Capra, The Tao Of Physics, 1989; p. 65.

6. JAMA-Archives 0f Internal Medicine, Vol- 159, No-19, 25th Oct, 1999; pp. 2273-2278

7. Strig Theory and Cosmology- How Old is the Universe?
www.superstringtheory.com/cosmo.html

8. Henry Gray, Gray's Anatomy, 35th Edition, 1973; p. 16

9. Robbins-PATHOLOGICAL BASIS OF DISEASE. 5th Edition; Neoplasia, 1994; pp. 257-272.

Materialization by photo identity of Almighty B.B.B By different scientists
1. Trinth Xuan Thuan- The Changing Universe –
   Big Bang and After-- Creation and Destruction by
                              Almighty B.B.B
2. Fritjof Capra - Tao of Phtsics - Creation and Destruction
                           of particles by Almighty B.B.B
3. Colin A. Ronan - Deep Space – Creation and Destruction
4. Depiction of Creation and Destruction by Almighty B.B.B
5. Viewing Home of Almighty B.B.B the creator and destroyer

Myself in left hand of my father and my cousin on right hand

www.ingramcontent.com/pod-product-compliance
Lightning Source LLC
Chambersburg PA
CBHW040828180526
45159CB00001B/104